AF260792

PYRAMID BUILDING

The "Big Bang"
For Science, Technology & Industrialization

DR. SAM O. OTUKOL

Library of Congress Control Number: 2022915006

HARDBACK: 978-1-957575-98-8
PAPERBACK: 978-1-957575-97-1
EBOOK: 978-1-957575-99-5

Ordering Information:

For orders and inquiries, please contact:
1-888-404-1388
www.goldtouchpress.com
book.orders@goldtouchpress.com

Printed in the United States of America

CONTENTS

Prologue For Pyramid Building Book ix

Chapter 1:   Introduction..................................... 1
Chapter 2:   Egypt is an African Country 19
Chapter 3:   Why The Super Square Method Worked .. 29
Chapter 4:   How Construction was accomplished 35
Chapter 5:   Why Super Square method worked........... 39
Chapter 6:   Why Was This Method,
              not Revealed Earlier? 49
Chapter 7:   The Pyramid Structural details 54
Chapter 8:   Support Structures 58
Chapter 9:   Can The Super Square Method
              be Verified? 64
Chapter 10:   Why is the Revelation of the Super
              Square Method Important? 68
Chapter 11:   Quarrying the stones 73
Chapter 12:   Significance of Pyramids to modern
              Civilization 77
Chapter 13:   The mathematics of pyramid design 82
Chapter 14:   Documenting the proof 87
Chapter 15:   Plan for a Documentary 91
Chapter 16:   Restoring the Khufu Pyramid 96

LIST OF FIGURES

Figure1: Map of Africa showing location of Egypt. 20
Figure2: Map of Egypt showing historic burial
grounds 21
Figure3: This is the out casing of Pharaoh
Tutankhamun's sarcophagus, it shows
the advancement of workmanship and
skills of cutting stone with unparalleled
precision. 26
Figure4: Damages to invaluable statues and
artifacts by European adventurers. 27
Figure5: Mutilation of facial features of
monuments by invading adventurers. 28
Figure6: A figurine showing a unique symbol
on the face. The symbol predates
the Christian era and had spiritual
significance. This symbol relates to the
inner square and the 4 side triangles in
Figure 12 29
Figure7: The facial symbol predates the
Christian era and had a different
meaning different from the cross. The
symbol on the face relates to the inner
square and the 4 side triangles in Figure 12... 39

Figure8:   Establishing the key bearings (East, West, South, North). 41

Figure9:   Defining the parameter of the outer square (the Super Square) 42

Figure10: Dividing the outer square sides into three segments and numbering the nodes (1 & 2)..................................... 43

Figure11: Defining the boundaries of the inner pyramid base square and the 4 corner isosceles triangles by connecting the nodes 1 & 2 as indicated below. 44

Figure12: The inner pyramid base square is defined when all the nodes are connected. Note the location of the isosceles triangles. 45

Figure13: The climate of Egypt changed dramatically over the millennia. It was an area with trees and other vegetation......... 61

Figure14: Geometric scaling of the pyramid in relation to earth dimensions......................... 83

Figure15: Aerial photo of the pyramids on the Giza Plateau. 87

PROLOGUE FOR PYRAMID BUILDING BOOK

For people who have been to Washington DC, or if you live there, you may have been to the Washington monument. You have probably gazed at it and were awed by the beauty, the rigidity, the elegance, and the distinctiveness of it. If it struck accord with you, you were probably not aware that you were gazing at an obelisk - an erect African male organ. To be more specific, you were gazing at the male organ of the Egyptian god Osiris.

Osiris had a lover, a goddess called Isis. According to Egyptian folklore, the two of them were madly in love, however, Osiris had enemies. One of those enemies kidnapped him, killed him, and dismembered his body. Then the body parts were scattered all over the world.

Isis missed Osiris very much and looked for him everywhere. She eventually learned that he had been killed and that he was dismembered and the body parts were scattered all over the world. She set an objective to find all his parts and reassemble him. She succeeded in doing just that. But to her shock and disappointment, his male part was missing!

Isis was not deterred. She used her divine powers and fabricated the organ. You could say, she met and exceeded the dream of every woman in the world – she had fabricated an organ that was permanently rock hard and available 24/7. Isis did not stop there, she used the fabricated organ to impregnate herself. After the usual nine months, she gave birth to a son they named Horace. They were gods so you cannot question why they gave the child the name Horace.

Anyway, from that time on, Egyptians began building obelisks. These are phallic symbols acknowledging the genius of Isis. Today, only one of the many that existed over millennia is still standing where it was originally placed. That one was commissioned by one of the few female Egyptian Pharaohs – Pharaoh Hatshepsut. She must have erected one as a show of power and strength because she ruled in a male-dominated world.

In Washington, you may have also noted that a short distance from the obelisk, there is a capital hill with a domed building housing the House of the Congress and government offices. According to the folklore of conspiracy mongers, the dome represents the womb of Isis. Why the Washington city planners chose to place the obelisk and the domed building in close proximity is a mystery. Perhaps some historians may explain why this is so, but that is another story.

It is worth noting, that the Vatican Holly See has a similar arrangement of dome and obelisk. In this case, it is not clear why a Holy place where pastors are committed to observing celibacy would adopt a dome and an obelisk as key architectural features.

The point of telling the background of obelisks is to illustrate how linked existing cultures are to our historic past as a human race. Many Americans quite often ask:

"What contributions have Africans really made to modern-day civilization?" Yet, right in the middle of Washington DC is an obelisk, which is a concept originating from Egyptian (African) history.

Almost 100% of the people who see the Washington monument don't see the linkage to African history and how significant that is to our modern-day living.

This book on pyramid building is an attempt to show how the execution of the construction gave birth to assembly line concepts and industrialization in general. It was the first visible evidence showing that a human civilization applied mathematics, physics, and advanced science to produce free-standing structures that have lasted for more than eight millennia.

CHAPTER 1

Introduction

Mention of the word *PYRAMID* to anyone in the twenty-first century invokes images from exotic historical places such as Egypt, Mexico, and Asia. The pyramids in Egypt and Mexico in particular invoke awe at the ingenuity and mathematical precision that was used to build them. Modern civilization is still struggling with questions of what tools were used to cut and transport the huge stones. How were people of that era able to lift stones weighing 200 tons or more to such high elevations?

A casual examination of ancient pyramids seems to bear limited significance to contemporary life. In modern times the focus is on the academic value of historic information. In reality, pyramids have shaped our very existence, in that they gave birth to science and technology. For instance, all sectors of mathematics, including arithmetic, geometry, trigonometry, calculus, and even building fractals were engaged in designing and building pyramids.

In fact, a pyramid is a summary statement of all relationships, including social, economic, political, family, but most significantly, metaphysical relationships within solar systems, galaxies, and the universe.

In a social relationship, there is an overwhelming drive by individuals to be the first, the best, the richest, the person with the best house. In economic terms, every agency, company, or department is led by one chief executive officer from whom all decisions emanate. In politics, people are driven to be the supreme leader of the lot in the county. In a solar system, the sun controls planets and asteroids within its zone of gravitational pull. In a galaxy, all solar systems conform to the range of influence of the center of the galaxy. Stars within a galaxy rarely peel off and seek their own direction. Within the universe, there is one guiding force that influences all galaxies. Galaxies do not routinely collide – they move in predetermined formation.

In a pyramid, there may be millions of stones, but they are all linked, and at the top, there is one that cups them all. In the structure, there is a state of harmony among the stones. But the pyramid is also linked with the stars within a galaxy and outside the galaxy. So, it could be said that a single pyramid is in harmony with all objects in the universe. That in fact, was the primary purpose of the pyramid – to link its occupants with earthly beings and the celestial objects in the universe.

Even in these modern times, there is no single government, company, agency, or society that functions coherently without a pyramid structure. All organizational charts contain the ever-pervasive structure. To avoid implementing the structure in an organization spells disaster in the long run.

The obsession of some ancient civilizations with pyramid building is instructive, but pyramid building did not necessarily, always produce positive results. In this book, it is shown that the practice actually accelerated the desertification of the Sahara zone. On the positive side, the practice led to techniques of organizing Human Resources to build complex structures for the betterment of human life.

Ancient African Concepts of Good and Evil

All cultures that exist today have had concepts of good and evil in their history that have shaped their contemporary spiritual values. In China for example, the contemporary values of the people there can be explained by what Confucius taught. He stated that moral values start at home. A family as a primary unit of a population in a country shape how people relate to each other. In a family, children learn how to respect their parents and elders. The children also learn to respect each other, and from that they learn that they should treat other siblings the way they expect the other siblings to treat them. These family lessons translate into the behaviour of how family members behave to other people in the community.

In Ancient African spiritual beliefs, the focus was on assessing the factors that influence the behaviour of individual community members toward other members in the community. In this regard, it was believed that behaviour was influenced by two spiritual forces; the trickster (an agent of chaos and evil) and the Devine (an agent of good and orderly behaviour) in society.

The Trickster force

If a person was under the extreme influence of the trickster, they would be more likely to promote chaos, war, excessive drinking, prolific in fornication, murder, violence, tardiness, laziness and anarchy. If you met such a person on the road, they would be unkempt, loud and vulgar. They would probably attack you without provocation. They would not be interested in your welfare.

People under the influence of the trickster are responsible for all the turmoil and tribulations in communities. They delight in the suffering of people, in violent confrontations, in wars, in premature deaths, in illicit activities, in orgies, in binge drinking, in adultery and in fornication.

The Devine force

The Devine force is the good influence. In its extreme, it represents, peace, harmony, order and perfect relationships among people in a community. Ardent followers of the Devine force are always seeking compromise and reconciliation among members of society.

If you met a person under the influence of the Devine, they would be well-groomed, well-dressed, and very civil. when they greet you, they would be genuinely interested in your welfare and would offer you assistance if you were in distress.

There is a sliding scale of influence of the trickster as opposed to the influence of the Devine force. There is a continuous and persistent effort by the Devine force to counter balance the influence of the trickster. On a daily basis, there is a constant battle by the trickster force to overwhelm and to overcome the influence of the Devine.

The choice of whether a person follows the trickster or the Devine depend depends purely on an individual.

A person who chooses to be totally and fully under the influence of the trickster makes decisions based totally on his free will. His or her actions are guided by the trickster, but may chose not to undertake actions that may result in criminal offences. A person who is totally under the influence of the Devine force, surrenders all his will and their actions are guided solely by the Devine force.

Understandably, the majority of the people in the community tend to play in the middle of the extremes. Following the trickster too closely my be dangerous to the health of the follower, and other people in the community. On the other had surrendering totally to the Devine means giving up all worldly interests and devoting ones life entirely to the Devine.

Ancient Egyptian spiritual values

The premedical trickster/Devine values formed the foundation of early Egyptian beliefs. More specifically, the Egyptians believed that the behaviour of a person while he is alive affected his/ her destiny when they died. Upon dying, the Powers of the underworld - Osiris, weighed the heart of the dead person. The heart of a very good person weighed less than the weight of a feather. However, the weight of the heart of an evil person would not pass the feather test. Such a person would endure numerous tribulations in the afterlife.

From the judgment suggestion you can see the sliding scale system embodied in the trickster/ Devine forces beliefs. This interpretation of good and evil exists to some extent in all religions in the world. It is the foundation of law and order for all governments in the world. Spiritual systems that differ from this model are nothing but witchcraft or satanism.

In modern day Christianity, the trickster is the devil or satan, and the Devine is God. As in ancient African spirituality, satan is playing the same role as the trickster. The difference is that satanism has become more organized and now has even places of congregation, and discipline is enforced even more rigorously that it is in Christian institutions. The Devine is better defined now than it was 10,000 year ago before the birth of Jesus Christ. At that time trinity was not really well established. The idea of God the Father, the Son and the Holly Spirit were not clear. Technically, the Son (Jesus Christ) was not defined 10,000 years ago.

The pharaoh - the head of the Egyptian state in ancient history was considered to be fathered by the sun. He was a God. For that reason he was granted exceptional treatment in life and in death by his subjects. When a Pharaoh died, he was transported straight into space to become a star. His body was interred, but his spirit was among the stars.

To distinguish the pharaoh from ordinary folk, preparation for his burial was distinct. At first, successive Pharaohs were buried at a common location under one mound. Subsequently, as the Egyptian society developed, burial arrangements become more sophisticated.

Contemporary concepts of good and evil

Over many millennia human spiritual values have not changed much. Both the trickster and the Devine are constantly struggling to occupy your mind. It does not matter what religion you may belong to, every hour, minute or second, your mind is invades by evil or Devine thoughts. What makes a good or evil person depends on which influence is dominant; the trickster or the devil. It does not matter what what level of social standing, level of education, amount of wealth, country one lives in or kind of job. Any person's mind can be highjacked by either the trickster or the Devine force. In Judaeo-Christian philosophy, it is written that as one thinks, so he or she is.

It is easy to conceal ones predominant evil thoughts and convey an image of a good person. There are many examples of this in life. In Canada a high ranking officer in the airforce fooled many people with his clean image of an outstanding officer. However, it was later discovered that he was a sexual predator and psychopath that had murdered several women.

The ancient Egyptians had a system for for spiritual assessment of individuals. And one had to rely on self discipline to avoid behaviours that might lead to eternal damnation.

For that reason Egyptian burial rituals were designed to ensure that an individual travelled safely through eternity without being overwhelmed by evil influences. The dignity with which one was conveyed to the final burial destination contributed significantly to their reaching their eternal destination.

Historical background of pyramid construction

The discussion of pyramids in this book will be limited to those in Egypt. For the author, having been born near the source of the Nile, provides a connection to the subject. It may not be known to many people, but there is evidence indicating that pyramids may have been built as far south as northern Uganda.

Progression to building advanced pyramids such as those existing at Giza was a culmination of many millennia of paying respect to family members who passed away. As far back as 10,000 BC, Egyptians experimented with various modifications of methods of preparing burial grounds for the dearly departed. Their obsession with ensuring comfort for the dead has influenced virtually all cultures on earth to date.

Before 6000 BC, the burial ground was a rectangular space called mastabas. Within the rectangle was a selection of valuable possessions of the dearly departed. The selection included household goods, pets, and in some cases domesticated animals such as cattle if the deceased was a cattle keeper. And within it, there was a grave where the deceased was laid.

For many millennia, the arrangements in the mastabas did not change much, except that wealthier people had more elaborate arrangements than the less wealthy. Then starting around 7000 BC they began experimenting with building mounds over the mastabas.

Initially, the mounds were just earth piled over the mastabas. For the wealthier individuals, consideration was given to access to the burial area. For instance, a wealthy person with a large extended family might want the closest family members who pass away after him or her to be buried close to him or her. This desire evolved into the construction of layered step pyramids made of laid bricks.

The first layer was created when the first primary family leader passed on. When a son or other close family members passed on, another layer was added, and so on.

Layered pyramids were cumbersome to build and, in some cases, they were failures to execute.

The step pyramids evolved into angled-sided pyramids. These also followed a layering process. Not too many angled-sided pyramids were built. They were awkward-looking and prone to collapsing.

It was not until approximately 6000 BC that the Giza-type pyramids began appearing on the landscape. More than 50 of these dangerous pyramids were built within Egypt. Perhaps there are some that were built and are lost in the desert sands.

The Giza-type pyramids were a wonder to look at but were disastrous in terms of environmental degradation. Their weakness was in the method of construction. This methodology will be described in considerable detail later in the book.

The reality about ancient history is that it is based on stringing together loose threads. Some conclusions from such loose threads have to be viewed cautiously. When you consider that some of the writers of history try to build the pyramid suggesting that they own the key to the truth, you have to screen the chaff from the wheat yourself.

Contemporary theories on construction methodologies

The practical efforts to pile two million tons of two-to-twenty ton rocks on top of each other are considerable. There is no doubt that it requires meticulous planning and rigorous personnel supervision to ensure success. The planners knew what challenges they were facing for the entire duration of the construction process.

Every step of the construction process was well thought out in advance. For us looking at the finished product, more than five millennia after the fact, the challenge is guessing accurately what happened when all the key players were long gone.

There are several theories suggesting how it may have been done. Three of these theories will be discussed. They include Lost Levitation Technology, The External Ramp, and Internal Ramp theories.

The Lost Levitation Technology Theory

In this methodology, the proponents suggest that the Egyptians of that time possessed a technology that they could harness to cause rock blocks to levitate and be moved through the air to the desired location. If such a technology did exist, it would have been wonderful. It would reduce work tremendously and be very helpful in speeding up work completion.

Imagine picking up a 10-ton rock, raising it high in the air, and placing it perfectly where you planned to place it within minutes. You would then move to the next rock and repeat the process. And you would keep on doing this all day without you the operator ever getting exhausted. This would be truly amazing.

If such a technology existed, it is unlikely that it would have been allowed to die out. The only reason such technology would be lost is if it involved secretive practices of sorcery or witchcraft. In these scenarios, only a few individuals would know what was involved. And if this is the case then it would not be a technology, but perhaps black magic.

The other limitation of the levitation technology is that it is not clear how high it could lift heavy rocks effectively. Could it lift rocks to 100 meters high? How exactly did it work? How big was the actuator that did the lifting? Was it working from ground level or was it placed above ground? How far was it from the base of the pyramid?

It seems that the levitation methodology raises more questions than provides answers. If in fact, it employed black magic, would that be consistent with the Devine aspirations of the Pharaoh who was the intended key resident of the building? Could it be just a fairy tale conjured up European visitors desperate to comprehend a puzzling historic wonder?

Although this construction methodology has credibility challenges, it deserves inclusion because it is not totally out of the realm of possibility. We can leave it as a theory and await proof of its efficacy. Who knows ... there might be a time when it might be revealed and proved through demonstration that it works? At that time, we can chalk it down as a theory.

The External Ramp Method

The external ramp method is the most popular among all. In this method, two-ton rocks are pulled up a rump until they reach the location of placement. Twenty to thirty strong, well-trained, healthy men are deployed to pull the rocks using ropes. The length of the ropes would be 10 to 20 meters from the face of the rock. A crew of men covered the rump surface with slippery substances to reduce friction between the rock bottom and the rump surface.

There were several conditions necessary to make the ramp methodology successful. Only a few are discussed here.

The most important condition was that the slope of the rump had to be less than 10%. The reasoning here is that the smaller the rump slope tilt, the easier it would be to pull the stones up the rump. Steeper slopes would exhaust the pulling crews too quickly requiring frequent stops along the way. Also, stopping on a steep slope might result in a rock rolling back to lower levels which would cause inefficiency in overall work.

The second condition for efficient use of a ramp was that the surface of the ramp had to be uniformly firm. If soft sections were encountered, the leading edge of rock would collapse the soil and get wedged such that forward movement would've been hindered. Such incidents would result in temporary work stoppage and time wastage. Considering rope tension had to be maintained once upward work started, any prolonged work stoppage was potentially disastrous.

The rump arrangement was usually outside the sides of the pyramid. For that reason, the construction was ring-like, spiraling higher and higher to the top of the pyramid. It had a corkscrew appearance.

For this method, it is easy to see how the first three layers of rocks would be assembled with ease. Beyond the third level of rocks, complications emerge. For the first layer of rocks, they could be dragged directly onto the desired location. For the second level of rocks, a suitable direct ramp can be built to supply rocks. An enhanced rump can supply rocks to the third level. A ramp would then start at the second level to supply rocks for the fourth.

At this point, two obstacles emerge. The first obstacle is that the slope of the pyramid itself was in most cases 51 degrees. Building a compacted earth rump around that slope and ensuring that the earth did not slide off, must have been time-consuming. How did they ensure that the compacted earth did not slide off the face of the pyramid?

The second obstacle was that as the ramp progressed from the second level to the fourth, the pulling crews had to negotiate corners around the rocks on the third level to get to the fourth. The question is, how did the builders do it? It seems then that work from four levels of rock to the top was cumbersome. Once again, we have to consider this method as a theory. It has never been proved as totally feasible for completing Khufu-type straight angle pyramids.

The third obstacle is that logically; one would need to build three to five levels of stone before building the ramp. If the ramp was built for each layer of rocks, there would be too many ramp loops, and they would be too fragile to bundle the traffic of many rocks being lifted.

The other obstacle is the possibility that rain would most likely damage the ramp frequently depending on the frequency of showers in the area. Furthermore, rainwater running down the side of the pyramid weakens the rump.

For this reason, the use of an external ramp would be inefficient and expensive because it would require frequent repairs.

The final possible obstacle is that the length of the ramp along each side becomes progressively shorter as you move to the top of the pyramid. That would make it necessary to increase the ramp slope as you approach the top. If this happened, moving rocks would be riskier as the workers approached the top. The work would become harder.

Overall, the ramp total length would go to fifteen kilometers long. You can imagine the difficulty of pulling one rock for fifteen kilometers. Then you have to haul two million rocks. In other words, the people pulling the rocks would be walking close to a total of 30 million kilometers over a period of 20 to 30 years that it took to complete a pyramid. This is an almost unbearable distance.

Many crews would be working shifts to do the work, but overall, it was a very considerable effort. But as you might conclude, dragging two-ton rocks up a rump is not the most efficient way of building a pyramid where more than a million huge rocks are involved. It would be a miracle to finish a pyramid in 20 years using this method.

The Internal Ramp Method

This method was proposed in 2003 by a French architect called Jean-Pierre Houdin. He spent ten years researching the feasibility of the method even before he visited one. He appealed to many Egyptologists to review his theory, but none were interested.

Eventually, he visited the United States of America where he found an Egyptologist who was willing to listen to him. He presented drawings of his theory and used his knowledge of architecture to convince the Americans to join him in the investigation. The two of them eventually visited Egypt to see real pyramids.

Jean-Pierre's theory is that the pyramids were built from the outside, starting with the limestone cover stones, to the inside. And to do that, inside rumps were used. It is not clear how the rumps were deployed, but the contention was that the workers crawled about inside the pyramid until they finally reached the top and placed the capping stone.

A concerted search of the evidence of internal rump did not produce concrete evidence of their being used. Jean-Pierre's partner was even granted to climb the outside of one of the Giza pyramids to search for clues. For some unexplained reason, Jean-Pierre himself was not granted permission to climb the outside of the pyramid, but he was given free access to internal chambers which he explored extensively.

The internal ramp method has its own limitations. For instance, how did the ramps negotiate around internal chambers within the pyramid? Once the walls began to rise, how were rocks moved around corners in the labyrinth of ramps in the pyramid? Was there one continuous internal ramp spiraling through the pyramid or were there two or more? At the end of the construction, how were the ramps filled in? And finally, how was the capping stone placed on top of the pyramid to finish the work off?

These are serious questions requiring clear answers. Overall, it seems this method raises more questions than provides answers. It is a method that is difficult to prove conclusively, and it may take time for it to transform from theory to fact. It remains a possibility, and it is up to the proponent to provide further evidence of proof.

The Rock-melting and Remolding Method

In this method, it is alleged that it is impossible to cut stones into blocks and fit them seamlessly as the Egyptians did. Instead, it is claimed that the Egyptians had solar lenses that were used to either liquefy or convert the rocks to powder.

This liquefied material was then poured into molds and re-solidified the mix back to rock.

It is true, that it is a bit tricky to cut rock to the exact sizes that are found in the pyramid construction. When the rocks are pushed together, they fit so well, no space is left between rocks. Also, all the rocks at a given level of the pyramid are exactly the same height, such that they form an even level.

However, we know that Egyptians could cut a tunnel through rock and maintain exact dimensions through the entire tunnel, or reduce the size of the tunnel systematically depending on the desired effect. This issue of exact sizing was not a big problem for the Egyptian builders.

Liquefying, rocks and re-solidifying them seems to introduce a new challenge to the process, and would probably slow the progress of the work. It increases the logistics required to handle the rock blocks, and the transportation of the powder or liquid rock from the ground level to the construction site is not clear.

A physics definition of work

In contemporary times, when someone says they are working, it does not necessarily mean that they are exerting themselves physically. It may mean a person is sitting at a desk answering emails or standing in front of an audience speaking. It may also mean shoveling sand into a wheelbarrow and moving it from one place to another. The shoveling sand job is called labor.

In Physics, there is an actual definition of the word work and it is associated with a means of quantification. In Physics, work is a measure of energy transfer that occurs when an object is moved over a distance by an external force. Now that is a mouthful, but it means applying energy to move an object from one place to another.

In the case of building a pyramid, it means applying energy to move a rock from one place to another, which means moving rock from outside the pyramid and placing it at a higher level at some point within the pyramid wall. It also means moving a rock along a ramp for several meters or kilometers depending on the desired destination. There is actually a mathematical equation that can be used to calculate the energy required to move a two-ton rock 100 meters on a 10% slope, but it is not presented here to simplify life.

If you have men pulling the rock, it is hard work. But if you can use horses, it is easier on the man's life. It is even easier if you can use a tow truck. Anything that can be deployed to simplify work can be called a machine. Except, if you are using a horse, it is simply a beast of burden.

The concept of work is explained here because it is clear that at the time of building pyramids, the Egyptians were fully aware of how to simplify work. They were the most advanced people in understanding Physics at that time. Europeans who have tried to explain the construction of pyramids have always discounted what the Egyptians knew at that time. Usually, assumptions are imposed on the situation. For instance, people may say the wheel was not invented at that time! Or they may say; iron had not been discovered by that time.

What we are dealing with is lost information. No one really knows what tools of scientific knowledge were available to them so they should be given the benefit of doubt. If we cannot reconstruct a pyramid today even with our advanced technology, why can't we accept the possibility, that the Egyptians might have possessed knowledge we don't yet have. Not that it did not exist, but simply that it existed, but was lost at some point.

The next new methodology, never described before, approached pyramid construction with an open mind. It relies on the possibility of finding clear physical evidence to tell a tale signs that a physical structure was built to make the construction possible.

CHAPTER 2

Egypt is an African Country

For some prejudiced people, it is difficult to accept that Egypt is in Africa. In reality, it is. Its location is unique. It claims the honor of occupying the final stages on the great Nile River before it empties its mineral-laden waters into the Mediterranean Sea. The Nile is Africa's longest river and on a global scale, it is the second-longest river in the world.

Figure 1. Map of Africa showing location of Egypt.

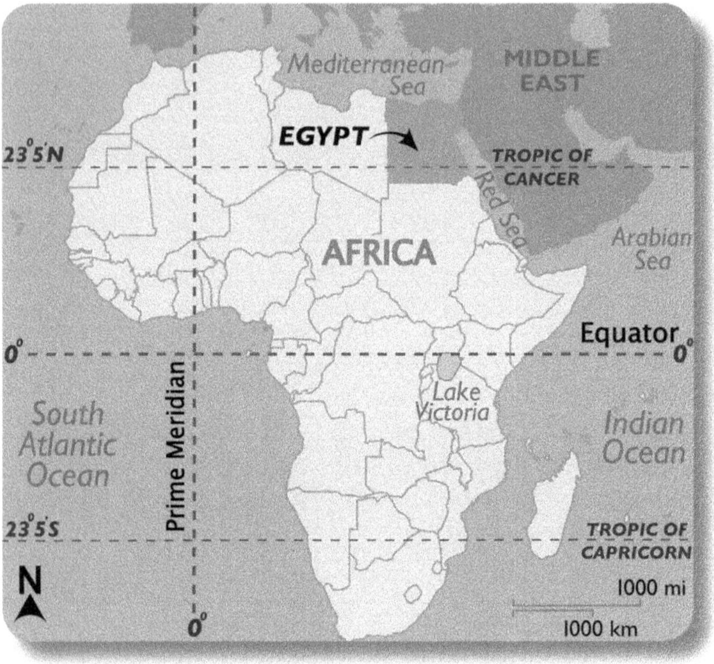

The second unique aspect of Egypt is that it is crossed by the Tropic of Cancer., which has considerable significance in that it gets the full impact of the solar solstice. On that point, the link to the spiritual aspects of the sun as the source of all life assumes immense meaning.

For Egypt all forces that drive life, i.e., water, sun, and fertile soils coverage and it is no wonder that miracles of nature and science happened here.

Figure 2. Map of Egypt showing historic burial grounds

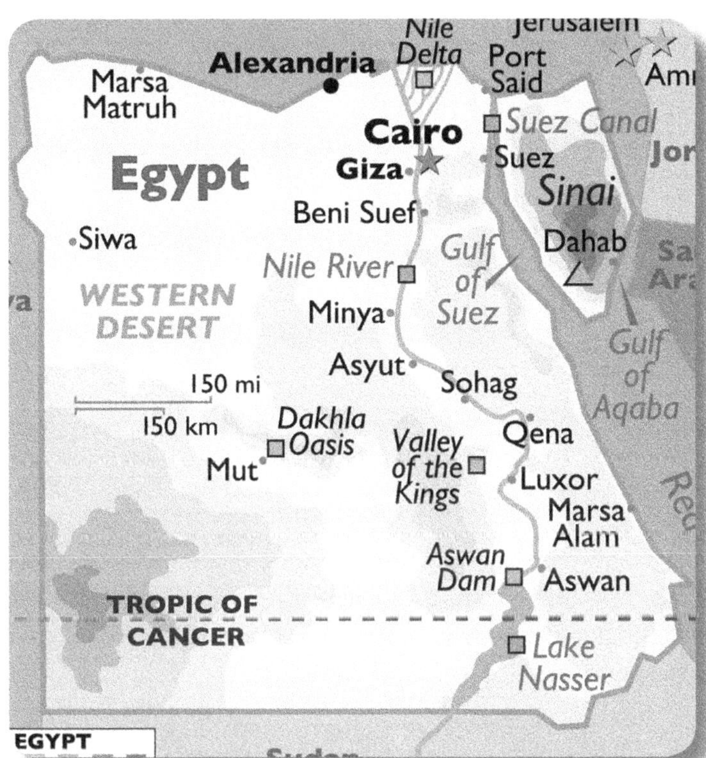

In general, you could say the thirtieth parallel cuts Egypt in half, with 50% of the land being below and 50% being above.

Location was an advantage for Egypt. If we assume the world 10 million years ago was cold or mostly frozen, it was a good place to be. The tilted earth's axis would work to its advantage during the time around the summer solstice. For the months of May, June, July, and August, this zone would be the warmest on the globe.

With this view in mind, Egypt would have been a perfect place for early human settlement. This advantage may have nurtured a progressive and innovative society. People may have settled in that area for many millennia before other organized societies piped up elsewhere.

The Egyptian education was legendary. Many luminaries such as Aristotle, Homer, and others, studied there. So, it is inevitable that there was a technology transfer between Egypt and Europe. Some concepts such as the mathematical value, the magic number, the Archimedes principle, and the Pythagoras theorem were probably well known in Egypt before they crossed over to Europe.

There are thousands of pyramids scattered in Egypt, but the most interesting ones are located in the Giza plateau, which happens to be quite close to present-day Cairo, the capital. The Giza pyramids are the primary attraction for pyramid enthusiasts. The Khafre, Khufu, and Menkaure pyramids are the most prominent.

Map 2 shows where the Giza plateau is, close to Cairo. However, the Valley of the Kings is the second most popular area for the burial of Egyptian royalty. It is located at Luxor in upper Egypt.

Both the Giza pyramids and the Luxor Valley of the Kings are located on the west side of the River Nile, indicating a place of death. The east side of the Nile was considered the place of life.

The most valuable agricultural land was located close to the Nile River. This was partly because these areas were flooded almost annually. With the floods came silt from many places up the Nile River, covering a huge catchment area stretching from the highlands of Ethiopia to as far south as the equator. This catchment area covers millions of square kilometers of land, silt from that vast area worked its way down to the Nile delta area. There is no doubt that the lower Nile basin was incredibly fertile. There are many biblical stories attesting to this. When other neighboring countries were in famine distress, Egypt seemed to have food to support them.

In spite of the fertility gift, it seems the lands surrounding Egypt still turned to desert. It is quite likely that the land stretching from the Nile west, to as far as the Atlantic may at one time have been lush with vegetation and teeming with wildlife. In fact, there are signs in the desert that big trees and lush vegetation once existed there. There are even cave paintings showing the existence of lakes and abundant game. But it all vanished. What was it that caused that catastrophic loss?

Based on the pyramid construct theory proposed in this book, there is no doubt that the pyramid's construction required vast amounts of timber to complete. The three great pyramids of the Giza plateau, in particular, consumed billions of cubic meters of timber to construct. The Egyptians must have hunted down every reasonably sized tree from thousands of miles around to build them.

It is interesting to note that after the great pyramids, none bigger or the same size was ever built again. In a much smaller one, which was built, and the last two were step pyramids. The question then is; how come the subsequent ones were smaller? Could it be that the timber resources were exhausted? Could it be that the Egyptians totally deforested the Sahara zone leaving it bare of any trees? Could this have accelerated desertification?

Some Egyptian studies indicate that there was a phase where severe famines happened. There are even hieroglyphs showing starving people. Could these famines have coincided with the beginning of desertification? Could it be that deforestation was the main reason Egyptians stopped burying Pharaohs in pyramids?

The Pharaohs were very powerful and forceful people. If a Pharaoh decided they preferred burial in a pyramid, there was no way an ordinary mortal would have stopped them. They could only be stopped by an immutable factor; no human being could control.

Being buried in a pyramid was a status symbol. It was like the difference between traveling through eternity in a luxury vehicle as opposed to traveling on a motorcycle. Being buried underground like commoners would have been a hard pill to swallow for a Pharaoh if they had better options. The Khufu, Khafre, and Menkaure pyramids had set a high bar, and every Pharaoh would have aspired to meet or exceed that bar within financial limits.

There is no sign to show that the Pharaohs who came after Khufu had become progressively poorer. For example, King Tut who died at an early age had already accumulated considerable gold assets in his early years. His predecessors would have had a similar fortune and would have been able to build better after-death accommodations. King Tut himself was buried in an improvised hurried grave in the Valley of Kings.

The relocation of burials to the Valley of Kings is a development that needs close examination. More specifically, did desertification have anything to do with it?

As a local superpower, Egypt dominated her neighbors.

Today, the neighboring countries are Sudan to the South, Libya to the West, Saudi Arabia to the East, Israel to the North-east, and the Mediterranean to the North. In ancient times, some of these neighbors did not exist as viable state entities. Those that existed were much weaker.

The state of Israel did not exist during pyramid-building times. It was a state that was conceived and lived in gustation in Egypt and was born at the time of the Moses exodus. Up until the time of the birth of Jesus, it was a state of minor consequence in the scheme of military and economic power. Saudi Arabia was for a long time a loose coalition of nomadic tribes that would have presented limited opposition to Egypt.

Today, Ethiopia does not have a common border with Egypt. But in ancient times, it was in the Egyptian sphere of influence and may have been required to pay the Egyptians dividends for peaceful coexistence. Sudan had fierce Nubian tribes that occasionally staged military challenges, but they too preferred peaceful coexistence most of the time.

To the West, the once productive zone deteriorated when desertification come. Here too, tribes were incoherent militarily and were largely nomadic. Egypt could have entered that area without opposition if it was necessary.

As it eventually turned out, the Mediterranean side was Egypt's weakest border. The Europeans who had been embroiled in their own local issues grew to admire Egypt and developed a hunger for her wealth. At first, the physical presence of the Mediterranean Sea was limiting, but eventually, the desire to acquire new territory become overwhelming.

In ancient times though, Egyptian technological advancement was light years ahead of that in Europe. When the Europeans arrived, they could not accept that ancient Egypt was an integrated society where black Africans contributed as much to Egyptian development, even more than the lighter-skinned Egyptians.

Figure 3. This is the out casing of Pharaoh Tutankhamun's sarcophagus, it shows the advancement of workmanship and skills of cutting stone with unparalleled precision.

The Europeans were for some reason frustrated by the fact that evidence of black contribution to the Egyptian civilization was everywhere. So, they went on a rampage trying to destroy evidence of black participation in Egyptian history. For instance, the nose of the Sphinx was blown off by a gun destroying an invaluable millennia-old monument. Many artifacts and statues showing negro features were mutilated.

Figure 4. Damages to invaluable statues and artifacts by European adventurers.

Explosives were planted in the Khufu pyramid in a primitive effort to knock it down. The attempt failed. These acts can only be described as barbaric. The European barbarians had arrived, and they were determined to take Egypt down. And that is where it is now.

Figure 5. Mutilation of facial features of monuments by invading adventurers.

The European hypocrisy regarding equality with black (negro) people should end. The two races are not in competition for the ability to create new things or comparison of intelligence. God has endowed both races with talents that they deploy for the development and enjoyment of all. The black people have produced great musicians. In fact, black people have indelibly affected American and world music to the point that without it, the world music landscape would be totally different. But black people are not begging for accolades for it.

CHAPTER 3

Why The Super Square Method Worked

Figure 6. A figurine showing a unique symbol on the face. The symbol predates the Christian era and had spiritual significance. This symbol relates to the inner square and the 4 side triangles in Figure 12

The construction of the pyramids was a spiritual exercise for the Pharaoh to spend eternity in a place of honor and comfort. In their time, there was no distinction between science and spiritualism. The Pharaoh was spiritually connected to God and the subjects recognized the Pharaoh's position as God's representation given in his area of jurisdiction.

Science was God's instructions to select people to create miracles that would glorify the Pharaoh. Spiritualism explained God's desire for his people to live a good life and provided instructions on dos and don'ts. Spiritualism explained the laws of nature. For instance, man and woman are different and will have an affinity for each other, but the relationship must be entered into mostly for procreation and companionship, not for gratuitous enjoyment. God also provided instructions on respecting the creations.

Today, some scientists believe God does not exist, and that they can interpret nature through studies. They see no link between science and spiritualism.

In pyramid construction, both science and spiritualism were pursued at the same time. For science, pyramid construction was the equivalent of the Big Bang, in relation to its application of science to human life. It is through pyramid construction that mathematics and physics evolved. It is reasonable to say, industrial production processes such as specialization, and fabrication of products through steps evolved. There were few other processed that required organized sourcing of raw materials and assembling them to create a complex product such as a pyramid.

The construction of a pyramid was linked to the rising and setting of the sun. For that matter, the drawing of the outline of the pyramid plan was guided by the rising sun.

This was particularly important for building the burial place of a Pharaoh, who was considered to be an offspring of a god and was in fact a god. It is fair to speculate that the starting of the pyramid construction was timed to coincide with the time of the solar equinox, as this was the time when the sun was close to the northern hemisphere. By picking a specific date, such as June 20, the orientation of the pyramids would be consistent from one to another through time.

The four triangles at the northeast, southeast, southwest, and northwest corners of the pyramid body provided a framework for establishing workstations from where independent crews could operate. The outer square also provided an opportunity for drawing a contour map of the pyramid, on papyrus paper. This allowed the calculation of the exact number of rocks required to complete a pyramid.

Being able to launch four crews at the same time made it complete the work in a reasonable time. If the total crew number was 100 people, each crew would consist of 25 individuals. As such personnel were utilized more effectively. One hundred people working at a single location would be too crowded.

Each of the 4 crews would have a similar organizational structure. They would have a supervisor and a few team leaders. Some of the crew would man the tower where rocks were lifted. Others would have floor duty - moving rocks from pyramid edge to a destination location.

There was a floor plan for each level of rocks. Most rocks were pre-identified with a specific location on the floor plan. For instance, rocks lining passageways within a pyramid were pre-identified and were brought in when construction at the passage was due.

Some of the larger pyramids may contain up to 5 million cut stones. First, the internal stones were laid. Afterward, the cover stones were placed over the internal stones. One theory proposed by Jean-Pierre, a French architect, suggests that external cover stones were placed first, but it is difficult to see how that could have been done.

The lift towers' elevators seem to be more defensible. They would reduce the amount of labor. In addition, they allow four crews to work simultaneously. One advantage that does not seem obvious, is the possibility of using beasts of burden such as horses, donkeys, and oxen.

If the target were to lift 5 million stones, that would be quite a challenge. Assuming the workers were engaged for six days a week, which is 312 days a year, a minimum of 200 rocks would have to be move per day by each of the four crews. This is on the assumption that four workstations are deployed every day.

If only one crew is deployed per day, the number of rocks to be moved per day would increase to 801.

With that volume of material required to be moved within strict time limits, the planners had to seek out the most efficient technology to move the heavy load. Elevators had an advantage over any other possibility.

Is it reasonable to assume that the subject rocks were loaded onto some kind of sled? The sled would facilitate easy loading on an elevator and offloading off the elevator. The sled would also make it possible to move the rock across the construction level to a specific target location where the rock would finally rest. Again, the aim was to move material as quickly as physically possible.

The possibility of one rock getting bogged down for hours was not acceptable. The possibility of delay of completion was possibly severely punished.

During that era, working on a pyramid building project was a big privilege. Not only were they provided with comfortable lodgings, but they were also well-fed. Not only that, but the builders were also respected within the community.

The results of the construction provoked awe and shocking beauty. An ordinary visitor passing by a new pyramid was compelled to think of the Pharaoh in terms of divine association. It was not something an ordinary citizen was expected to produce. As such, is associated with being a builder or laborer in that endeavor invoked respect and elevation in society.

Today, a little bit more information is available on the people who built the pyramids. In 1922, an excavation unearthed a residential area within the Valley of Kings area, where it is believed the workers were housed.

The excavation revealed a lot of information about what life was like for them. It is said that they were well-fed, had supplies of beer and beef, and bread was available to them. They also had barbershops and areas where they could shop for food and other domestic goods.

It seems clear that they did not live in slavery conditions. They enjoyed life to a reasonable extent. Their working hours may have been long and some assignments may have been challenging. For instance, people positioning Roche at a destined location did less work than those pulling rocks for long distances. However, a more advanced level of skills was required to precisely position a rock than to pull it from general location A to another general location B.

As would be expected, the crews had managers and supervisors. Supervisors had better dwelling places than the average laborer.

Based on the excavated artifacts, the workers were also able to write letters, doodle for fun, and express grievances. Whether grievances were actually submitted as official complaints is not clear.

Some of the excavated writing includes love letters and romantic poems. So, the workers had time for personal reflections, pursued hobbies, and explored dreams.

Most important of all, the excavations revealed pottery and other utensils used in day-to-day life. Most of these items were found in areas where waste was dumped. It is these items that were most revealing. For instance, they revealed that the workers drank beer. Also, in the refuse dumps were fragments of bones and from these, it was possible to determine which animals the workers were eating.

CHAPTER 4

How Construction was accomplished

Most Egyptologists who try to explain the pyramid construction phenomenon start with the negative side. They start by saying things like … this and that had not been invented by that time, or that technology did not start until such and such time.

In this book, we avoid that negativity, and treat the builders as highly skilled engineers and use logic to reach conclusions. Evidence shows that the engineers found ways to move millions of multi-ton rocks to high elevations and assemble them with incredible accuracy.

We can see the stones in place and the structures are still standing even after a period of several millennia. Obviously, this was not work done by a bunch of drunken idiots doing things half-haphazardly. They were fully aware of the challenges they were facing and figured out solutions accordingly.

The current bigoted western civilization experts, interpret the pyramid construction from the current state of advancement of the African people. And sometimes even resort to insulting the pyramid builders by claiming that a more advanced civilization arrived from some unknown place and built the pyramids.

The interpretation of the Egyptian advancement is compounded by the fact that at the time of pyramid construction, European communities still lived in caves. And for all that matters, the Europeans of that time believed the earth was flat. It is something the Egyptian scientists never believed in as far back as 10 millennia ago. So now we are forced to believe, the Africans could never have been ahead of the Europeans at any time in history. It is ridiculous thinking.

We live in a cruel world. Even with no physical evidence, some so-called Egyptologists have concluded that pyramids must have been built by aliens from out of space. The question is where are those aliens now? Why would they spend so much time building wonderful structures then disappear without a trace? It does not make sense.

Another group of people has claimed the Pyramids were built by Enoch - a Jewish prophet who had much favor with God. There are so many issues with this theory, that it is difficult to even start an argument against it.

Had all pyramids been built within one century, there would be a remote possibility that it could have happened. But clearly, the pyramids were built over a period of many millennia. So the possibility of Enoch's involvement in pyramid building is non-existent.

To these African people haters, it is too painful to accept that Africans were at one time more advanced than any other race. What we should be asking is how did the Africans lose the advanced knowledge? Could this tragic set of circumstances occur to another civilization? What can humanity do to avoid a repetition of this information tragedy?

With modern data storage technology, it might be possible to archive information about our state of scientific advancement. But you should not bet on it. Egypt had an excellent library at Alexandria, but it was mysteriously burnt to the ground. All the records in the library were destroyed.

It is quite possible that the library may have been burnt by a deranged arsonist. Such is the barbaric nature of our humanity. This cycle of destroying controversial historic information has been repeated many times in other places.

The truth is that the Egyptians were the most advanced physicists and Engineers at the time when the pyramids were built. The world should accept this as a fact.

Earlier in this book, the physics definition of work was provided. To reiterate, it is the energy required to move an object of known weight for a known distance.

From this definition, it would have been clear to the Egyptians that moving heavy rock blocks up a ramp to any point at a high elevation was a non-starter. At a maximum slope of 10 degrees for a ramp, the minimum length of a ramp to move rocks up a 150-meter elevation would be at least 1.5 kilometers.

At this distance, and assuming only one ramp was wrapped around the pyramid, moving one rock alone to an elevation of 5 meters would take several days. For the bigger pyramids such as the one built by Khufu, more than 5 million rocks are lifted. Ramp method construction would have taken 50 to 80 years to complete. The 50 to 80-year timeframe is an entire generation. Very few Pharaohs would have been able to complete a pyramid in their lifetime. But history records indicate, most pyramids were built within a period of 20 to 25 years.

The 20 to 25-year construction timeframe was what a Pharaoh would have expected. After all, they wanted to finish their pyramid before they died. So the Egyptian engineers had to devise efficient means of moving rocks to meet their time constraints.

In chapter 5, a method for consistently orienting the pyramid so that the sides face the cardinal directions; North, East, South, and West, is described. One feature of the resulting construct is the four equilateral triangles. At first glance, these triangles seem to be redundant. But in fact, they are the big secret behind the pyramid's construction methodology.

Over these triangles, construction towers were built. These towers were in fact elevators. In these towers, rocks could be moved vertically, which is the shortest distance for moving any rock to its destination within the pyramid body.

Going back to the definition of work, elevators would be the most efficient means of reducing construction work. We believe the Egyptian engineers knew this. Now some critics would say … but pulley systems had not been invented by that time. Well, we suggest that the Egyptians must have invented them for that purpose. They were the best physicist at that time and were as interested in solving the work problem the same way we would be today.

This firm belief is supported by the fact that some of the rocks that were moved to higher elevations were long chamber ceilings which would have been impossible to move by ramp. The tower areas on the other hand were big enough to facilitate the movement of long pieces.

The use of the towers as elevators when confirmed as we surely can, will once again demonstrate the ingenuity of the Egyptian people. Their work was truly inspiring. It is just unfortunate that modern civilization has sought to degrade it for political and racial prejudice reasons.

It is perhaps a good idea to take a closer look at the Egyptian civilization of 5 to 7 millennia ago. We might be missing something significant and useful to our lives today.

CHAPTER 5

Why Super Square method worked

Figure 7. The facial symbol predates the Christian era and had a different meaning different from the cross. The symbol on the face relates to the inner square and the 4 side triangles in Figure 12.

The key to the success of the Egyptian construction methodologies was consistency and rigorous application of mathematical principles in the surveying and construction exercise. Also, they had traditions that had become rituals that they followed religiously. For instance, if the tradition dictated that the work had to start after the first sighting of the moon in a specific month, any other date would not be acceptable.

In this case, the establishment of the pyramid foundation plan had to follow a specific sequence of steps. The work had to start in June, on the day of the solar solstice. Special significance was placed on the solar solstice as it related to the day when the sun was closest to the earth. It was the day when the Father of the Pharaoh was most present. The Pharaoh was supposed to be conceived through the mating of the sun and the mother of the Pharaoh.

By using the same day of the same month over the years, there was a linkage of all pyramids to the same recurring event in the year. It reduced the wobble effect that would occur if the day and month changed for each pyramid.

In the pyramid East to West orientation, it was important to use the rising sun in the east as the primary bearing. Work on establishing the outer square diagonal could be conducted for most of the morning before the sun got too hot. And in the evening, at sunset, a back bearing could be taken to confirm the west direction of the diagonal (see Figures 8).

Figure 8. Establishing the key bearings (East, West, South, North).

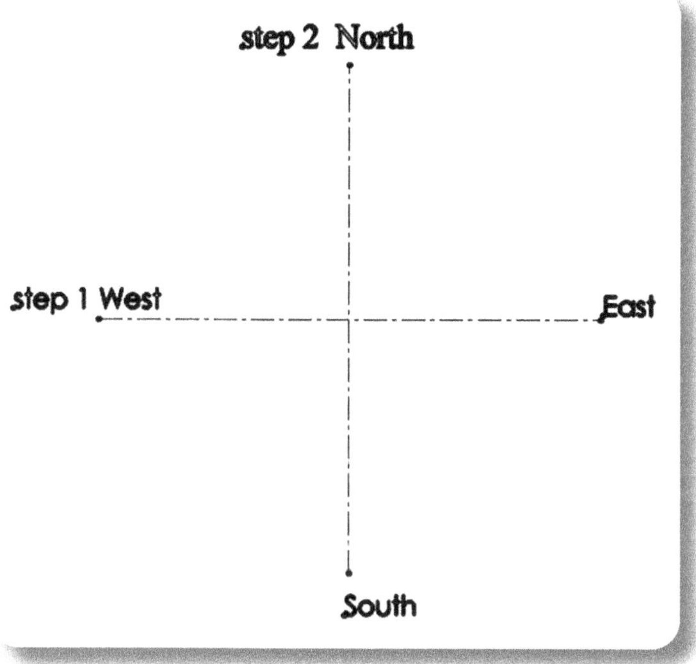

Figure 9. Defining the parameter of the outer square (the Super Square)

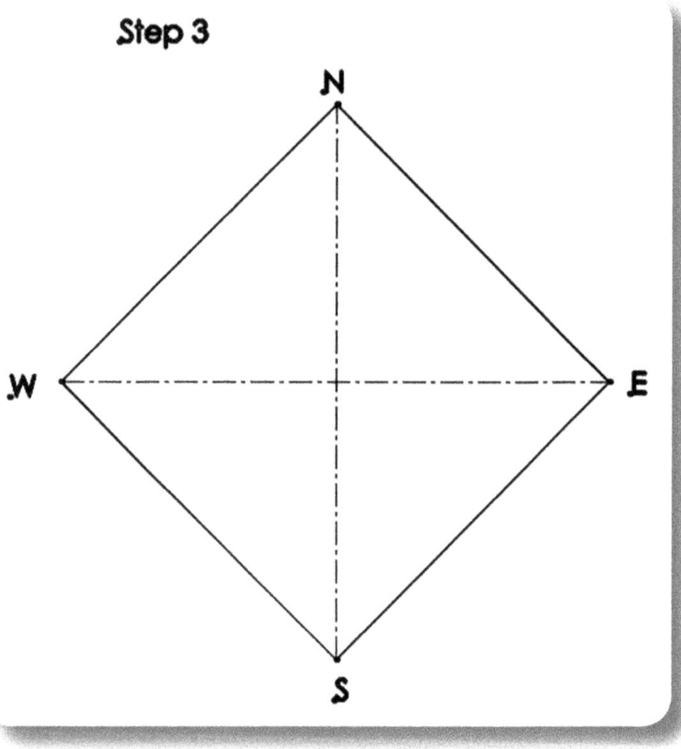

Figure 10. Dividing the outer square sides into three segments and numbering the nodes (1 & 2).

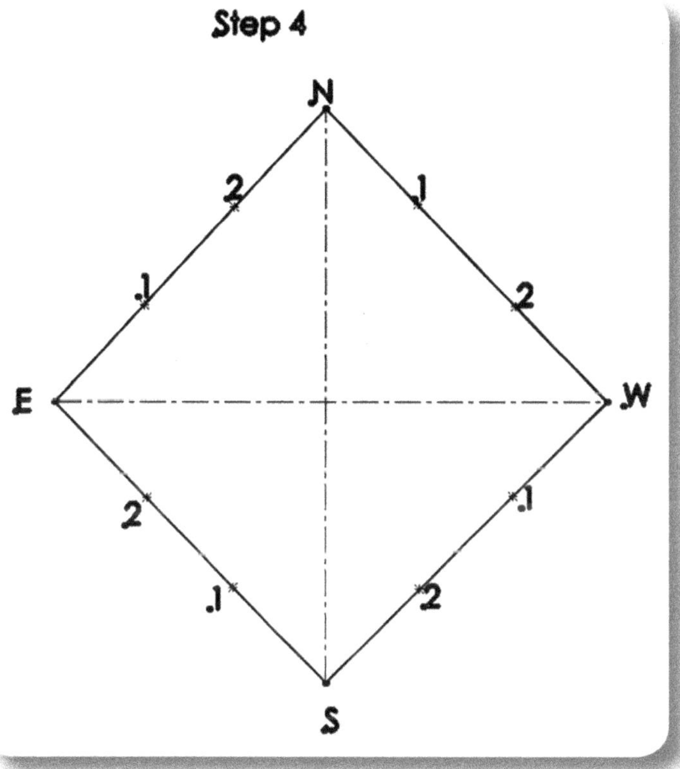

Figure 11. Defining the boundaries of the inner pyramid base square and the 4 corner isosceles triangles by connecting the nodes 1 & 2 as indicated below.

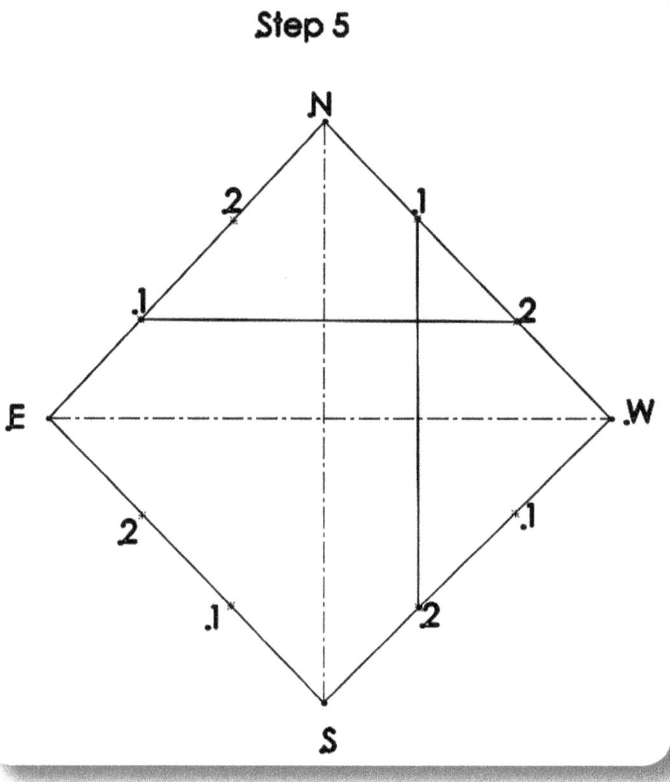

Figure 12. The inner pyramid base square is defined when all the nodes are connected. Note the location of the isosceles triangles.

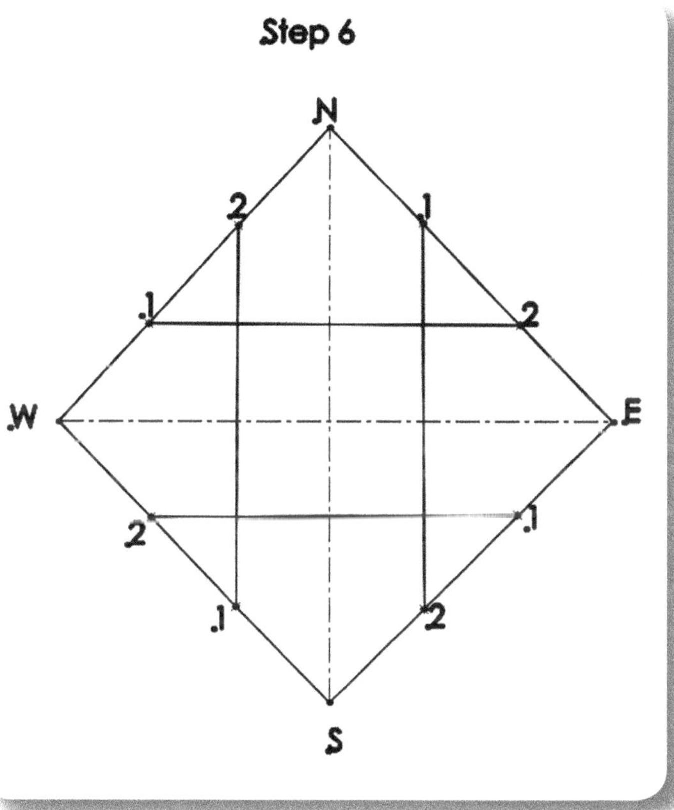

At the center of the East to West diagonal, a perpendicular line forming the North to South diagonal was established. The Egyptians had well-established methods for creating right angles between two lines and were able to establish squares with uncanny accuracy. It is quite likely that they were using what later came to be the Pythagoras theorem to confirm expected measurement in a square.

The Pythagoras theorem may have been in use for many millennia before Pythagoras provided the modern world with its mathematical proof. It is important to note that Pythagoras, did not discover the theorem. He only provided mathematical proof of it. This is clearly indicated in the article by the Editors of the Encyclopedia Britannica in the article "Pythagorean Theorem mathematics". The same authors also seems to indicate that Pythagoras went to Egypt for further studies before he produced proof of the theorem.

Once the outer square was defined, a process for defining the pyramid outline started. This process started with the dividing of the outer square into three exactly equal segments. Then the segment nodes were numbered as indicated in Figure 9 (see above). The nodes were numbered 1 and 2. In step 5, the number 1 was connected to the number 2 as indicated in Figure 11. The end result of the connections is shown in Figure 12.

You can see there is an inner square within the outer square. Each of the sides of that square faces a cardinal direction, North, East, South, and West. And as you can see, the process is quite simple and effective. The only tricky part is determining the right day to execute the orientation. But it seems that the Egyptians had quite accurate calendars and could apply other anecdotal local information to confirm the date.

If this super square method can be applied with some rigor, consistent results are produced each time. The important things to remember are that timing the rising sun on a specific date of a specific month was critical to maintaining consistency. Secondly, some mathematical principles were used to check and cross-check distances on squares. To that end, we could conclude that they may have used what we know now as the Pythagoras theorem to confirm and verify distances.

In the vertical direction, the builders constructed linear passages that were intended to be aligned with some stars in the sky. How they calculated angles to aim at specific stars will probably remain a mystery for many more years. In today's world, these scientists would be considered to be geniuses.

In Figure 12, you can see four isosceles triangles at each of the corners of the inner square. The exact lines of these triangles were very important. If the stone elevation towers were built there, stability was essential. These are the locations that bore the brunt of lifting the stones. Any weakness in the tower structures was intolerable. Any failure of one tower would delay work for many days or weeks or months. That was unacceptable.

The narrow side of the triangle was particularly important. All rocks passed through an opening on that side; it was the side of the tower that received the most ware and tare. The wood construction at that end of the tower was heavily reinforced for stability.

The side of the triangle facing the pyramid and the side opposite to it were open to allow the rock blocks to be loaded onto the tower from one end and offloaded at the other end. The adjacent sides were heavily reinforced but had holes through which the lifting popes passed to execute the lifting. The ropes were pulled in opposing directions at an angle perpendicular to the entrance and exit axis of the 'elevation' system.

A pulley system was used to move the stone blocks upward. Each cycle of lifting was about two cubits or nearly one-meter-long and special stops would be placed to rest the block before moving to the next level. This arrangement greatly reduced the amount of work required to move a rock block to any upper level from the ground. Rather than move blocks thousands of meters along a ramp, the distance was reduced to less than 200 meters vertically up an elevator tower.

It is important to understand that the builders were very advanced engineers, and they would not have missed the mechanical advantage of using an elevation system. It would have been insane to use any other means given the tight timelines that a pyramid construction required. A Pharaoh did not want to die before completing his or her pyramid. They wanted to be assured of a burial place before they died. It was an honor to be able to tour one's burial place and specify the desired decor before the final date came.

Clearly, the ramp method would require considerable time to complete a pyramid. For large pyramids such as those in the Giza plateau, more than 40 years would be required to finish a pyramid using the ramp method. The other methods including the rock liquefaction method involve additional steps that would prolong the work.

In the rock liquefaction method, first, the rock would be extracted from the quarry, then a solar-powered lens system would be used to liquefy the rock, and finally, the liquefied rock would be poured into moulds to form the rock block. Overall, there is too much movement of Material from place to place. From quarry to smelting site, from there to onsite mould formation.

CHAPTER 6

Why Was This Method, not Revealed Earlier?

If you visited New York City and went to the Empire State Building, standing outside you would clearly see that it is a very tall building. The building has 102 floors. If you ventured inside and went to the 100th floor, and looked outside, you would see that you are high up in the air.

If you happen to be coming from a small tribe in the Amazon, you would wonder; how on earth did these builders get building materials to the then 100th floor? Shit! It looks impossible to make such a building.

Similarly, if you went to St. Peter's Cathedral and saw the paintings on the ceiling and were told that a guy called Michelangelo did the painting. You might spend quite a few hours scratching your head wondering how he could have done it.

The reason you would be wondering about how the Empire State Building construction and the St. Peter's ceiling painting were accomplished is that the support scaffoldings were removed shortly after construction or painting.

The same is true for pyramid construction. The construction support structures were removed as soon as the work was declared as finished. In the case of the pyramids, the construction occurred many millennia ago, and there is no record of the skill sets that existed at the time of construction.

Most Egyptologists extrapolate based on known history or speculate based on their prejudices. The reality is that there is no evidentiary information to confirm a pyramid construction methodology. Well, not until now.

You are lucky to be reading this now because when the presented evidence is checked out and confirmed to be true, you will be one of the first people to know how it was done.

The other reason for finding limited or no information on the technique used to create pyramids is that the actual planning work for the construction was limited to a very small number of spiritually enlightened experts with specialized training. Very few people actually knew what those experts were doing.

The honor of doing the job was not available to everyone. The candidates had to be well-known individuals with a considerable reputation. In addition, they probably underwent additional training to ensure that they understood what was at stake and that they could follow standard protocols of the job.

And because this was a spiritual venture, they were probably prohibited from doing certain human activities while they were on the planning. If the planning component took two years to complete, they were probably prohibited from having sex until the job was done. I used sex abstention as a humorous example to illustrate the seriousness of their work.

The Pharaoh would definitely not be amused if he thought his chief tomb designer was busy fantasizing about orgasms while he was designing his afterlife residence. Such thoughts might nullify the security measures the spirits would accord him to safely navigate the dangerous journey of his future.

So the planning work for building a pyramid was undertaken under wraps and high security. The work included "food" plans, location of passageways, dimensions of stones to be placed at specific locations, measurements of dimensions of chambers, location of exit shafts, location of Pharaoh's burial chamber, locations of accompanying property chambers, location of key support staff, etc.

The planners had to be thoroughly familiar with all those customs and guidelines. Failure to follow guidelines could lead to unpleasant consequences. In one example, the builder miscalculated the height of the pyramid he was building. When the mistake was discovered after construction, the Pharaoh ordered the extra height above the pyramid to be removed. It is difficult to imagine what punishment the designer must have faced. It was not a fun job - designing a pyramid. But when it was done well the designer was accorded much favor by the Pharaoh, and the fellow subjects revered him not to mention that his family might get an extra free kilogram of the meat if they went grocery shopping.

After the construction of the pyramid was completed, the scaffolding and towers were not only removed but the holes that were dug to build the support structures were filled and possibly paved over. Some of the supports for the construction infrastructure were so well secured, they drilled into the underlying bedrock. These drilled holes in particular are what will prove the super square theory being explained in this book. The holes are there for sure. We just have to find them. They are located in predictable places, particularly in the equilateral triangle areas shown in Figure 12.

It is possible that the papyrus paper floor plans for the pyramid may have been preserved for posterity. If they were, they must have deteriorated over time. It is not a very durable material.

But on the other hand, the planners feared providing grave robbers with clear plans of the pyramid layout. For that reason, we should consider the possibility that the papyrus plans were destroyed.

It is unlikely that one Pharaoh would have wanted his predecessor to copy his tomb plans. It would have been bad luck to do so. After all, each Pharaoh needed to follow their own path to eternity. Copying another Pharaoh's pyramid plans would have been like seeking to follow his footpath. That would have been unacceptable.

In some cases, some Pharaohs were loathed by their predecessors. In such cases, the possibility of a loss of information is possible. For example, Pharaoh Khufu, who built what is considered to be the best pyramid in Egypt, has not had large statues of himself anywhere in Egypt. The only statue of him that exists today is a miserable little thing, less than one meter tall.

Other Pharaohs such as Ramses have larger-than-life statues several meters tall. Such Pharaohs literally spent their entire time on the throne planning to leave their DNA and footprint behind for others to see. You could call them megalomaniacs, but they gave us plenty of inspiration for thinking big. They made it possible that there can be no doubt that something significant and phenomenal happened in Africa.

In a way, they foresaw the future and realized that no one would believe that Africans could pull off such ingenious works. In this respect, they were right. No modern person today, including Africans, can believe Africa had that level of technology earlier than any other people on earth.

Most African leaders are not inspired to outdo the early Egyptians. Instead, we see African leaders beating their own people down so they can appear superior to them all. They do not strive to improve anything, but rather plan to accumulate more for themselves at the expense of their own people. It is a deplorable character trait that can only invoke derision and disgust.

How can we change this?

CHAPTER 7

The Pyramid Structural details

In modern building construction, every building must have a structural plan. The architect must first verify from the building owner what type of building they want. Is it a bungalow, back split, side split of two-story? Is it an apartment in a multi-unit building? Is it a detached or semidetached house? Does the owner need basement space and if so, what types of rooms should be there?

All these details must be provided to the architect even before they turn over any ground to start construction. The architect takes the suggestions and draws up a number of plans. Some of the plans will show how the house will look like when it is finished. Some of the plans will show the layout of the house showing where the different rooms will be, and room dimensions.

During pyramid construction, the same process of defining the three-dimensional internal space was followed. The super square space was helpful in making scale drawings of the pyramid. This exercise was essential to estimate how many stones were required to complete the structure. The estimation is necessary to avoid the possibility of underestimating the required number of stones.

Generally, the rocks at the bottom of the pyramid were larger than those close to the top. A contour map would have made a systematic reduction of the stone sizes possible. So, during the quarry work, the stones were cut to suit the level of the pyramid being targeted. Coordination was necessary to ensure work occurred according to schedule.

At each level of stone placement, open space for hallways, rooms, passages entrances, and exits were accounted for. Once the construction was done, there was no tunneling to create space of establish passages. Every stone stayed where it was placed permanently.

To achieve all this accuracy, planning had to be exact and concise. There was limited tolerance for errors. Major errors could weaken the entire structure and jeopardize the project. For that reason, the work of supervisors was recognized as being crucial to success. Consequently, the supervisors were accorded considerable respect.

Every space that was not occupied by a stone was well planned for and had a purpose. The most important chamber of course was the Pharaoh's burial room. Clearly, it was the largest - with ample space for the Pharaoh to relax in. Its dimension probably had numerological meaning; it had high ceilings. The entrance to the chamber was sealed after burial in order to reduce the chances of robbery afterward. From this chamber, there was a narrow passageway oriented in alignment with a specific star in the sky. The Pharaoh's coffin was placed in such a way that his propped up penis was in alignment with this passageway

The purpose of this alignment was to facilitate a mating opportunity between the star and the Pharaoh. You have to remember that in folklore, the Pharaoh was conceived by the sun mating with his mother. That is why he was considered to be a god. So a Pharaoh mating with a star was fair game. It was the way of the Egyptians. There was a requirement for him to have a direct line to shoot straight at the start. Whether or not he scored was another matter beyond human control.

Next to the Pharaoh's burial room were rooms housing his favorite objects. If he played the flute, the burial arrangement ensured that it was easily accessible. Then there were rooms where his favorite pages or helpers were housed. The helpers might include a concubine or two, or three. You'll never know when the Pharaoh might need a message.

In Pharaoh Tutankhamun's tomb, archeologists found chariot wheels. He either had a favorite chariot he did not want to part with, or he wanted to ensure history recorded that they had wheels back then. He foresaw that Europeans and people from other parts of the world might not believe he had chariots. There are some ignorant historians who up to now believe ancient Egyptians had no access to wheels.

All passageways were built to specification, and all had a purpose. The passages above ground were built as space unoccupied by rocks. Some passages that were built below ground had to be dug out of bedrock and had to be broken to create them. The one skill the Egyptians excelled in beyond all was digging or chipping through the stone. They worked stone almost the same way we work with wood. If they wanted to cut a stone of a specific size and shape, they did so with ease, or at least it seemed easy for them to do.

Once again, we come across the application of technology that contemporary Egyptologists dismiss as being unavailable in pyramid building times. Today we have laser technology and diamond toothed or diamond coated cutters which many believe did not exist back seven millennia ago. But the truth is in the pooping … if we see with our own eyes a product in physical form which we conclude needed advanced knowledge, we have to give credit where credit is due. Otherwise, we begin to delve into black magic and silly unscientific ideas.

I will discuss the technology idea further, later in the book. At this point, it is reasonable to assume that diamond-coated tools may have been available at that time. During the particular pyramid-building period, the Egyptians cut more stones than any other civilization. Even compared to current times, very few countries cut as much stone each year compared to the Egypt of the pyramid age.

In addition to cutting pyramid stones, they were shaping sculptures, obelisks, paving entire streets in stone, building houses of stone. The entire Stone Age is dominated by their work. Most of all many wonderful carvings of Pharaohs, gods, and symbols of power such as the Sphinx were cut out of the colossal stone of epic size.

CHAPTER 8

Support Structures

Of all the structures required to facilitate the construction of pyramids, the most important were the four corner towers located at the northwestern, southwestern, southeastern, and northeastern corners of the main pyramid body. These towers played a key role in simplifying the construction work so it could be completed within the twenty- to thirty-year period.

The tower base must have been dug very deep into the ground to ensure sturdiness and stability. Considerable effort must have been invested in ensuring that the completed setup would not fail under the heavy loads they would be subjected to during the construction process.

The towers were probably the most expensive aspect of the pyramid construction. The owners used an equilateral triangle formation originating from the sides of the larger square and terminating at the four corners of the pyramid. Because of its orientation, it had three clear sides from which operations could occur without interfering with the sides of the pyramid. Horses, donkeys, and human beings could work away on three sides pulling ropes anchored on pulleys allowing movement of rocks from the ground to any desired elevation, offloading the rock and lowering the platform to pick up another rock.

Several rocks could be moved at each tower location per day. The beauty of the tower was that both human and animal power could be deployed to move the rocks. In addition, the operation would have provided minimum risk to all participants both animal and human. The lifting was by outward motion away from the tower, which means that even when ropes broke and a rock fell, the pullers were moved from the tower walls.

The tower had a movable lift platform, which could move securely in the vertical direction (up and down) but was secured in the frame of the tower. At rest, the lift platform top was at level with the ground. A rock to be lifted was loaded onto a sled and pushed onto the platform.

The platform was attached to a system of ropes connected to pulleys. When the elevator was loaded, pulling the ropes would raise the platform together with the sled and rock load.

Based on the flexibility of using various lifting power sources, this was the most efficient method of building a pyramid. Otherwise, the workload would be too much. Imagine moving five million two-to-five ton rocks to elevations as high as 450feet (180 meters). Even very well-fed strong men could not do that lifting within a 20-year period without the assistance of beasts of burden.

In this case, oxen, donkeys, and horses could be deployed to assist in lifting the rocks. Humans could be deployed as well but in a reduced capacity.

Between the towers along the sides of the pyramid, scaffolding was built to assist in placing stones that were close to the edge of the pyramid. But perhaps the most important role of scaffolding was in the placement of the cover stones.

The dimensions of the Khufu pyramid are; width and length, 230 meters (780 feet) and height are 146 meters (481 feet). The footprint of the pyramid is larger than that of a football field. You can imagine the whole pyramid being surrounded by scaffolding. Considerable material was required to form the scaffolding.

The towers and the scaffolding were linked for stability and ease of movement of workers from tower to scaffolding and from one end of the pyramid to the other.

The towers were built from the ground to the highest point in one cycle. The side scaffolding on the other hand was built by the level of height of assembled rocks as the construction progressed.

When the laying of the cover stones started, work started at the top and progressed to the bottom. The scaffolding was dismantled at each rock level completion.

Timber was extensively used to build both the towers and scaffolding. This was the biggest disadvantage of building a pyramid because vast quantities of timber were required. Suitable timber for the building was difficult to find. This was particularly urgent for tower buildings. Very sound timber was required there. Timber was sought from far and wide. Constructing the three biggest pyramids may have virtually depleted all the timber in neighboring countries. After those three pyramids, subsequent pyramids were much smaller. In fact, the last two pyramids were step pyramids instead of straight side pyramids. Little or no timber was required to build step pyramids.

The interesting question is: did the excessive use of timber to build pyramids trigger desertification in countries surrounding Egypt? There may be some truth to this assertion. There was a period in Egyptian history when there was a prolonged and severe famine. Could those periods of famine have coincided with forest cover depletion?

Figure 13. The climate of Egypt changed dramatically over the millennia. It was an area with trees and other vegetation

This severe timber depletion did not affect Egypt alone. All countries neighboring it were affected. If in fact, Egypt needed the timber, no neighboring country would have dared to deny them access if they had it. It would have been taken by force if necessary.

It is not a new phenomenon that some power may pursue a national interest so diligently and blindly that they ignore warning signs of dangerous collateral damage.

Today we have similar situations where developed countries are indulging in the economic desertification of poorer countries. What May be looking like fair trade deals may be inflicting irreversible economic damage to weaker countries, which in the long run many never recover from poverty. Such countries are doomed to starvation civil wars and ultimately total destruction. The historic neighbors of Egypt are a stark example of the hidden consequences of insane policies.

The most significant possible effect of deforestation and subsequent desertification of the Sahara and the Middle Eastern area was that first, it resulted in a loss of all lakes that existed in the zone, and secondly, the released water vapor was blown over the Atlantic to North America. At that time that area was blanketed by ice which was as deep as one kilometer. The dramatic change in heat dynamics may have triggered the glacier melting era in North America.

Locally, deforestation resulted in less rainfall and frequent drought. With the drought, there was difficulty in growing food crops and subsequently, severe food shortages became the norm. Millions of people starved to death.

Unfortunately, because of a lack of secure data archiving, humanity lost the lesson learned from the events that followed deforestation. The citizens of that time could never have stopped deforestation. The forces seeking timber were too powerful to be challenged, the people were doomed to a terrible fate.

Such forces still exist today regarding various unjust practices. For instance, some politicians find ways to justify deforestation of the Amazon, and they do everything possible to achieve their goals. They even go to the extent of murdering those opposed to their insane deforestation plans. They displace indigenous populations, pollute rivers, and destroy wildlife sanctuaries all in the name of economic progress.

Similarly, developed countries go to poor countries and extract natural resources at ridiculously low prices, then bring the raw materials to their countries, fabricate goods from the materials, and sell finished products back to the poor countries at insanely high prices. It is all done in the name of what is euphemistically called free markets. How can poor countries get out of poverty if they sell their only possessions (raw materials) at low prices and buy finished products at very high prices? They are operating at a continuous loss! But the powerful don't really care about such lop-sided economics.

CHAPTER 9

Can The Super Square Method be Verified?

The problem with theories regarding how pyramids were built is that they are difficult to prove. For instance, the ramp method has never been reproduced by building a small pyramid that way. The levitation method is impossible to even attempt, because the device used to cause heavy rock levitation has never been recreated, so the method cannot be verified.

For the super square method, there are several locations outside the pyramids that can be checked to verify ground disturbance activities that we believe exist. For instance, the corners of the super square can be re-established. And at those corners, there may be signs of drilled holes, particularly if bedrock was penetrated.

In addition, the locations of the four equilateral triangles might be relocatable. Once again, these locations would be easier to find if the bedrock was disturbed. If the bedrock was drilled, the holes are still there., they cannot be obliterated.

At each of the four ground level corners of the pyramid, there should be deep holes going right into the bedrock indicating the outline of the tower bases. The foundations of the towers should have numerous drilled holes. The bases of the towers had to be firmly grounded for them to withstand twenty years of punishing heavy lifting work.

Any failure of a tower was a high risk to the timely completion of the pyramid. For that reason, no effort was spared to ensure that they would be durable and reliable.

The other reason why construction using towers was effective is that, even if one tower failed or collapsed, work would still continue through the remaining three. The challenge to the engineers was to ensure that all four towers operated at full capacity.

The towers were inspected regularly to detect weaknesses, and if any were detected repairs were implemented immediately.

All the construction work converged on to the towers. For that reason, there would be clear evidence of heavy traffic around them. If the traffic rolled over bedrock, there should be signs of scratch marks.

Although there has been a long passage of time, it might still be possible to find historic artifacts of lost equipment or Brocken pottery, that might confirm the presence of the original rock elevators. For instance, if a wooden plunk were placed in drilled rock, there may be holes where stumps of timber were stuck. Such stumps would provide a wealth of information. They would be easier to date and would allow the accurate determination of the exact date of pyramid construction. Such information is rare and would allow us to improve history records regarding when some pyramids were built.

It has been indicated earlier that four independent crews were used to construct a pyramid simultaneously. A study of the wear and tear at each of the four corners would refute or confirm this activity.

In addition to the towers' existence's evidence, the parameter of the base of the pyramid should have evidence of placement of support posts for the scaffolding that surrounded the entire pyramid during the duration of the pyramid's construction. The placement of the support posts would have been at regular intervals of two to five meters. Regularity would ensure the sturdiness of the scaffolding system and uniform distribution of weight stresses.

The other important factor of this methodology is proving the existence of the super square itself. There are twelve locations on the parameter of that square that can potentially be verified. The first one is the east bearing facing the sun. That one should be located at a distance equivalent to the width of the pyramid (which is 146 meters, 481ft) from the midpoint of the face of the base of the pyramid. This distance can be established exactly within one meter of accuracy.

The second verifiable location is exactly due west from the west side of the pyramid. The third and fourth locations are due north and due southern from the north and south faces of the pyramid the same 146-meter distance.

Each of the four sides of the super square was divided into three equal segments. On each side of the super square, two locations marking the ends of the mid-segment should be relocatable. Searches can be conducted to verify these locations. Their location should be easy because, if the length of the side is known, the locations should be one-third of that distance apart within a very narrow error tolerance of one meter or less.

The final verifiable aspect is to check for the alignment of some points of one side of the super square with a pyramid side parameter and an adjacent point on a super square parameter (see Figure 12).

If the expected alignments occur, and the tower footing are found, it will be a major breakthrough in finally establishing the method used to build pyramids. Establishing a method of construction is important to humanity as a whole and to Egypt in particular.

It is a crucial point in history that we need to know because it will help us to move forward together as people with a common past and get together to plan a future where we will all be considered contributors. Hopefully, this knowledge can unite us into a common purpose.

First, we should realize that the knowledge we are using today originated from further back in our history than we have been willing to accept. And that we should not claim ownership of technologies or science, because it is present in all of us and we should share it.

Our human tendency to marginalize less knowledgeable people is counterproductive. Their misery makes the so called more advanced people look harsh and unkind and reflects badly on everybody.

In the case of Africa where most of the knowledge started, it may seem that things there might never change. But as surely as the Europeans and other cultures learned from Africa, Africans can over time reverse the trend. If that were to occur, would the Africans be justified to bully people from other parts of the world? Absolutely not.

CHAPTER 10

Why is the Revelation of the Super Square Method Important?

For humanity, for nations, and for individuals, it is important to know our history. It informs us of where we have been and to some extent inspires us to be better people. For example, during World War II, Germany waged war against the other European countries. Japan waged war against other Asian countries and the United States of America. But in the end peace agreements were signed and all warring countries started to rebuild their lives.

There is a good record of the WWII events. Nobody in their right minds would want to replay that painful history. There is too much at stake. This lesson in history is helpful to our current generation only because, the history was well documented. It is something referred to as a reminder of the pain.

Today, even the most well-educated people are not aware of the role the African people had in originating scientific thinking. Most people are not aware that Africa had the tallest high-rise building; the pyramids, for more than five millennia without a challenge by any other continent. In fact, one might venture to say, it would have taken a long time for other cultures to build high-rise buildings had they not seen that it was not impossible by seeing the pyramids standing there.

Establishing the methods that the Egyptians used to build pyramids makes it possible to repair the existing crumbling structures. Without the repair, the existing pyramid structures will disappear within the next 5000 years. We should not let that happen. They are our most important heritage; not just for Africa, but for the world as a whole.

The current theory suggests that they may have been built by maybe up to four independent crews, operating from the four corners of a pyramid. If repair were to occur, a similar strategy would be the best way to go. At some point, humanity should commit to repairing at least one of the three biggest pyramids in the Giza plateau.

Telling this story and proving its veracity will prove to all who may be interested that Africans have ingenuity. In time the Africans themselves will know what they have to do to find their way out of the mess they are in. Already an increasing number are rising up from insecurity to prominence.

More and more are excelling in the arts, sciences, medicine, engineering, and business. They have reached a point-of-no-return and will change Africa and perhaps even the world. They have overcome colonialism; they will overcome neo-colonialism and finally dictatorship. A few enlightened business people have now identified that excelling in business is key to total emancipation. They have a keen audience and the ground is fertile. There is no turning them back no matter how manipulative the opponents may be. They cannot stop success.

Pyramid construction is an African heritage. In time, it will be proved that the Egyptians did get to Mexico many millennia before even the Europeans did. It will show that as bad as the idea might sound, very early on, Africans too were colonialists. They were probably as ruthless and inhuman when they were at it. But they lost that place in history and have taken forward new opportunities rather than look backward.

The continent has many challenges though. Close to 50% of the land area of the continent is either desert or severely arid. In order to manage the rapidly growing population of the continent, some of the desert and arid lands must be reclaimed. The same ingenuity that inspired pyramid construction can be deployed to solve desertification.

The persistence of berating Africans, by even educated people who should know better must be resisted by positive actions. Simple things such as simply accepting or acknowledging events such as the fact Jesus, the Holiest person in the Christian world, took refuge in Africa when his life was in Danger should be discussed frequently.

In addition, when there was a serious drought in Judaea and the Jews took refuge in Egypt, no other country could have been so accommodating. So why is more attention focused on the exodus that occurred 400 years later? Those Jews who were alive during the exodus would not have existed if it were not for the kindness of the Egyptians. Why do Christians fail to put these things in context?

Egypt was the incubator for the Jews that created the nation of Israel. Nobody should forget that notwithstanding the unfavorable condition they found themselves in after 400 years of exile. The Jews had enjoyed some good times, including land ownership and free access to the Pharaoh's house. Moses himself was raised by the Pharaoh's sister. He had access to the best education and unparalleled privileges.

Where else in the world would a person from a slave family have had such privileges? The Egyptians, and indeed most Africans in history and at present are kind and generous people. They tend to bend over backward to welcome aliens in their midst. But because of their gullibility, foreigners tend to take advantage of them.

Africans need not change from being kind and generous to others. However, they need to draw a line in the sand. They should define what values they will not compromise on. The first area is to get a lock on valuable natural resources.

Business people sell goods to earn income for their families. A business person who is not earning a profit from their business does not survive. All efforts should be made to ensure that raw materials being exported are producing enough return to make the lives of the originating country comfortable.

A revolution is necessary to make this change possible because the reasons for the failure of the resource exploitation economics is partly due to bad domestic politics and corruption. The African people should work collectively to cure themselves of bad politicians and root out corruption. Solutions to corruption will not come from outside. They must be handled from inside Africa.

The rising number of unemployed African youth will make revolution possible. Almost an entire generation of people in African will grow up without employment. This situation cannot stand. The youth must take action to ensure their own survival. The older people are tired, and most have achieved what they set out to get. Nothing should be expected of older people.

Instinctively, people seek peace and stability as they get older. There is a mentality of "don't rock the boat" among older people. These people also tend to use their past as a reference point. They see contemporary life as being better than what they experienced growing up. For them, they don't see issues such as corruption, unemployment, and suffering as big issues. After all, they too suffered during their youth. But the current circumstances are different. The resources are there and opportunities are plentiful, but mismanagement and white-collar theft are the problems. If the youth don't take action to correct these problems, they will suffer for a long time. They should not take this lying down.

CHAPTER 11

Quarrying the stones

The stones that were used to build the pyramids were not locally sourced. In fact, it seems that the most suitable rocks for building pyramids in the Giza plateau may have been quarried from as far south as 500 kilometers away.

The builders faced two challenges which were: a) to cut stones to customized sizes to a specific pyramid, and b) to transport the rocks to the building site in a timely fashion.

For each of the big pyramids in the Giza plateau, two to five million rock blocks were required. For that reason, the quarrying had to be fast and consistent.

For the quarrying method, we should not waste time debating whether they had or did not have the modern tools we have today? The point is that they did what to us seems impossible. They produced the stone blocks. My personal speculation is that they had diamond-coated tools. We use those today; one may well ask the question - how did we discover these tools if the information was not already floating in the ether.

To assume the Egyptians of that time could not possibly invent sophisticated tools is an insult to all Africans. Why deny them the boasting rights for a game-changing innovation they came up with before our time?

The stone cutting happened in narrow, hot places and required considerable effort. The stone cutters of that time should be commended for the incredible work they did.

Clearly, the Egyptians had honed their skills in stone cutting to an unparalleled level. They could curve stone as easily as we do wood in our contemporary technology. Examples of stoneware finished to exquisite standards abound. Not only that, Egyptians could drill deep holes into rock with as much ease as we do with a powered auger digging through the soil in today's world. Digging 5 or 10-meter holes through bedrock was not uncommon. And this was done without dynamite blasts. We have to give credit where credit is due. The builders were excellent at what they did.

It seems that cut stone blocks were moved to the River Nile, and from there floated toward the Giza plateau. Once again, we must acknowledge that the Egyptians must have known about the Archimedes principle before the Europeans came along to claim it as their own. The rock shippers had to correctly assess how much weight could be loaded onto a specific barge so as to avoid loss of load on the way to the destination. That they were successful at delivering their load to the construction sites consistently attests to this assertion.

The movement of the rock blocks to the river must have been done using sleds. Otherwise, it would not have been possible to move square-edged two-ton rocks by dragging on uneven bare ground. Once again innovation was employed here.

It is possible that the movement of stone blocks on dry land, was by horse, donkey, or oxen power. Harnessing human efforts would be too time-consuming. The one thing the builders could not afford was time wastage. Things had to be done in a timely manner.

The construction timelines were quite tight because when pyramid construction was happening, it was the highest priority project for a Pharaoh's administration.

The movement of stone blocks from the Nile ports to the pyramid site was once again by horse and sled. The work was synchronized to avoid the storage of too much backlog at the construction site.

The rock blocks at the bottom of the pyramid were generally larger than those at higher levels. For that reason, the delivery of rock blocks was coordinated, such that the right size of rock blocks was delivered at any given level of the pyramid construction.

At the quarry, work had to be organized such that the builders at the construction site provided instructions regarding the level they were working on. They gave instructions on which section of the pyramid they were needing blocks for. And if any specialty blocks were required to line a passageway, for instance, the stone had to be produced for what was required. In most cases, changes in rock size were announced several days or weeks in advance.

To maintain balance within the pyramid, all rocks at a given level were the same height. What was required was for the tops of all rocks at a given level to be even and flat. That way rock blocks sat on each other without leaving gaps due to surface unevenness.

As you can imagine, this was highly precise work, requiring ingenuity and focus. In Latin America, some ancient stone cutters shaped rocks to overlapped seamlessly - sometimes wrapping one rock around another. And that leaves one to wonder how much time was required to custom fit rocks over each other without leaving gaps.

It is a skill our current generation of stone cutters have virtually lost.

One has to also wonder if rock blocks were polished after the initial cut. Furthermore, one wonders if additional tweaks were made to the rock to ensure all surfaces were level or perpendicular.

All these sophisticated requirements to build a sound pyramid invoke admiration for the craftsmen and women who dedicated their lives to this endeavor. Less time should be dedicated to questioning what tools were used and devote more time on how we could do it today if we were required to do it. It is not something we would want to do as a matter of routine. It is something that involves too much physical exertion and concentration. It would be much costly to do in our modern economy, where labor is measured in time units.

One thing is for sure; those ancient stone cutters and builders deserve respect and honor. They left us a legacy we might have missed. In that view, some thought should be given to restoring at least one of the great pyramids in the Giza plateau for posterity and as a means of preserving these works for another 10,000 years. A monument to the ingenuity of the Egyptians should be preserved for future generations.

Some cultures are restoring buildings constructed 300 hundred years ago. Why should we not consider restoring some pyramids that have stood for over 4000 years?

CHAPTER 12

Significance of Pyramids to modern Civilization

There is a mystery around pyramids that our modern civilization has yet to grasp. There is a message the builders were trying to convey that we are glossing over. It is the aspect of the pyramid as a model for harmony. What was harmony all about? Was it an attempt to link earth with interstellar cosmic forces? Was it a model for building better relationships among human beings? Was it a message from spiritual forces against discord among people of different ethnic, tribal, and racial backgrounds here on earth? Can we learn anything about this monumental endeavor? Or have we totally missed the point?

In 1633, a mathematician called Galileo Galilei was tried by the Catholic Church for supporting an idea by Copernicus which proposed that the earth revolved around the sun. At that time, the church understood that the earth was in a fixed place and the sun revolved around it. It is a long interesting story. The point of the story is that back then science and religion were not clearly separated.

Incidentally, at the same time, Europeans believed that the earth was flat. The idea that a person in Alaska was standing above someone at the equator was unacceptable.

This point is very important because as will be explained below, the Egyptians of ten thousand years earlier knew the earth was round. And were well ahead of Galileo's assertion that the earth revolved around the sun.

A presentation by Professor Randall Carlson reveals something about pyramids that is mind bugling. Five thousand years before Jesus was born, the Egyptians had highly advanced mathematical information about the earth and were using some of that information to design pyramids.

The height of the Khufu pyramid is 482 feet and the diagonal distance from the center of the pyramid to any of the corners is: 502 feet? The really surprising thing about the measurements is that when you multiply the pyramid height by the number 43,200, you get the radius of the earth from the center of the earth to the North Pole. And when you multiply the same number with the diagonal distance, you get the radius of the earth at the equator. And guess what, 43,200 is the number of seconds in half a day (12 hours). Obviously, considerable knowledge of the dimensions of the earth was required to execute such an extrapolation.

What that means is that by design, the Khufu pyramid was a reduced model of the Northern hemisphere of the earth. These measurements were checked by satellite survey, which is the most accurate way of doing it, and were found to be within 300 feet of error. This is totally mind bugling. It means, Egyptians had the ability to take incredibly accurate measurements of the size of the earth 10,000 years ago!

Professor Randall also found that for a period of two seconds, the earth at the equator rotates a distance equal to the base width or length of the pyramid. These numbers have been verified scientifically.

This information shows that the Egyptians were light years ahead of their time in terms of mathematical knowledge. It also confirms our belief that questioning how they did this or that is futile. What level of technological advancement they had is more than what we can handle.

Clearly, the Egyptians were the first people to apply mathematics to solve practical problems. Even the fact that they could stack millions of rocks together, and include open spaces in the midst of the rocks is a miraculous wonder. They had to balance the weights of the rocks well enough to avoid internal rock collapse.

The solutions they applied back then are applied today as a matter of routine, but the science had to be developed. It must have taken time to get things perfect, but the Egyptians were suited to the task. Their results are there to be witnessed to date.

The fact that the Egyptians scaled the pyramids to earth dimensions goes back to the concept of harmony. Why would they go to the extent of scaling pyramids to earth dimensions?

It may be that the Egyptians were trying to create a cosmic connection between the Pharaoh, the pyramid, the solar system, our galaxy, and the entire universe. Incidentally, Galileo did not know galaxies existed. He just assumed that outside the solar system there was just the universe.

Oddly, it seems the Egyptians were conducting their exploration of the solar system, and the universe without the aid of telescopes.

The linkage between pyramid and harmony is something we should explore more deeply. Is it possible that we could find solutions to our current social, religious, racial, ethnic, and economic strife.?

Clearly, strife abounds even in most expected groupings such as religion. Ever since the creation of Christianity, religion has fragmented into smaller and smaller sects, some of which are not even coherent. Christians and Muslims don't see eye to eye. Religious intolerance is the norm. It has reached the extent that religion has become the top source of hate between people.

Among Christians, Protestants hate Catholics, Baptists despise all other sects, and Pentecostalism believes they will be the only ones that will make it to heaven.

The only religion that practices a semblance of peacefulness is Buddhism. But even they may have an inherent dislike for some other grouping. Judaism has no respect for Jesus who in fact was one of their own.

The white people would like to possess the blacks as slaves in some form or another. The majority of them would care less if black people lived under a bridge in filth far away from their own homes.

We cannot build decent societies if we nurture all these latent prejudices. We can continue living a lie or embrace the concept of harmony and start caring about one another. In Christianity the one thing Jesus talked about the most was hypocrisy. This is the situation where people pretend to be on one side when the opposite is the case.

In Egypt, it seems that an effort was made to balance things. It was at one point, a place where a stranger could go and feel at home. This was evident when at the time Jesus was born, his parents chose to go there rather than go with the wise men from the east. In spite of the story of enslavement, and the tragic exodus, Jesus was safer in Egypt than anywhere else.

It is strange, that the present-day Christians largely ignore the role Egypt played in saving the life of infant Jesus. This failure to acknowledge such small incidents is typical of the deep levels of mistrust and sometimes even hatred communities harbor against others who may have never committed any offense. Hatred is pervasive. It permeates all levels of society and communities. There is an urgent need to try different social experiments.

CHAPTER 13

The mathematics of pyramid design

The best reference on the mysteries of pyramid design is a lecture by professor Randall Carlson. Professor Carlson has considerable expertise in numerology and the mysteries of ancient societies. He presents an incredible picture of how advanced the Egyptians were in the knowledge of the dimensions of the earth.

For example, he provides the following basic numbers which you should keep in mind as we continue the discussion.

Number of seconds in 1 minute. 1 x 60.　　　　= 60
Number of seconds in one hour. 60 x 60　　　= 3600
Of seconds in 12 hours.　　　　12 x 60 x 60 = 43,200
Of seconds in 24 hours.　　　　24 x 60 x 60 = 46,400

There are other numbers he believes recur in interesting patterns, but we will stick with these for argument's sake. Please check out his presentation in this YouTube link if you want to know more.

The number to watch for in this presentation is 43,200, the number of seconds in twelve hours.

For the Khufu pyramid, if you multiply the height (482.8) of the pyramid by 43,200, you get the radius of the earth from the center of the earth to the North Pole. This calculation is 482.8 x 43,200 = 3,949.834miles. The actual satellite-based measurement is 3,949.8934 miles. The difference is 313 feet.

Centre

Considering that the Khufu pyramid was built more than 4000 years ago and the satellite measurement occurred in 1982, what are the chances that the two estimates would be that close? This means the science the Egyptians were using to build the pyramid was quite advanced!

Figure 14. Geometric scaling of the pyramid in relation to earth dimensions

Conversely, if you take the diagonal distance from the Centre of the pyramid to the corner on the pyramid, which is: 484.37, and multiply it by 43,200, you get the radius of the earth at the equator, which is 3,963 miles. Here too you can see incredible accuracy of prediction of the radius of the earth.

The amazing thing is this magic number is the number of seconds in 12 hours (half a day). Obviously, the pyramid designers were linking the earth's characteristics with the Pharaoh's place of final rest. And on a grander scale linking time and space with the final destiny (eternity) of the Pharaoh. It was a work of genius.

To achieve all these design specifications, the pyramid designers had to be highly educated people. They had to have had advanced training in mathematics, physics, astronomy, and spiritual history and traditions.

Some of the required knowledge was tribal - passed on from generation to generation. For instance, using the rising sun on a specific day of the year and the means for orienting the pyramid must have been a tradition passed from generation to generation. We should assume that June 20 or 21 was the date they were using. This is the solar solstice the longest daylight of the year. This would make sense since the conception of the Pharaoh was linked to the participation of the sun. The solar solstice is a day of honor for the sun and links well with the construction of the Pharaoh's final place aboard.

Modern-day engineers and construction workers would be curious to know what tools the workers of that time were using. It is a relevant interest. However, the emphasis should not be on the known evolution of technology, the evidence is in the results. In fact, the big challenge is that the burden of proof is on us to prove the ancient engineers wrong. They skillfully assembled the rocks. Let us prove we can do better with today's technology.

Clearly, today's technology can produce similar results. But what price would we be willing to pay to do it? Alternatively, if we cannot match what they did let's pay homage to their ingenuity. And teach it to our children that smarter people than us once lived here and technological advancement is not necessarily additive or linear. What we see is possible and the Egyptian engineers did it.

Let's not mislead our children by making it seem that it was the work of witchcraft or black magic. The pyramids were built by real people like us. It is just unfortunate, that their technology was temporarily lost. But the fact that we can do the same things today and perhaps, even more, advanced things must be due to our ability to observe problems and create solutions.

In addition to developing technology to solve problems, Egyptians had also advanced in terms of human resource management. Imagine the effort required to recruit a large number of staff, assembling them in various locations and coordinating their activities such that they work as one team to produce a breath-taking work of art at a colossal scale.

It was indeed a monumental feat. Not only were large crews housed, fed, and perhaps even entertained for many years covering the duration of the pyramid construction, tight schedules were met as planned.

Today many projects are started every day, but a few failures are due to a lack of coordination. The more complicated a project may be, the greater the chance that it may fail. This did not happen often with pyramid construction.

Among the Egyptian pyramids, the capping stone of one of the pyramids was not installed. The pyramid was completed as was expected, however, a review indicated that the height had been miscalculated. The patron of the project - the Pharaoh ordered the top to be removed.

This demonstrates the seriousness with which accuracy was taken in pyramid construction. No Pharaoh would have accepted to lie in a poorly finished tomb. There were standards that had to be met.

The managers and supervisors of pyramid construction projects lived under a lot of stress. They had to keep their staff well-motivated, but be stern enough to meet milestones. The administration had to work like a well-oiled machine. Delays and downtime were to be reduced to a minimum.

Work was allocated according to skill and specialization. Again, our modern society may have benefitted from seeing the impossible being achieved in pyramid construction.

Some workers were separated into divisions. For instance, if we accept that four separate crews were needed to build four corners of the pyramid, then one manager controlled each corner work station. But on top of that, one overall site manager was needed, to ensure that work progressed at an equal pace from all four corners.

Work on all four corners was finished before moving to the next level of the pyramid. If work at any of the corners was behind schedule, workers were reassigned to join the deficient corner to balance the work and finish all four at the same time.

CHAPTER 14

Documenting the proof

The claim of the use of the super square method is only a theory at this point? In order for it to be accepted as fact, we will need proof. The challenge is to enroll sponsors and well-wishers to make proof a reality. There is a lot at stake here. Most proposed theories so far have not been proved. The reason is that the scale of the work is too Honorius. For our theory though, we do not have to climb all over the existing pyramids to find evidence. We think the evidence is outside and around the pyramid. It is drilled into the surrounding bedrock where no one has ever looked before.

Figure 15: Aerial photo of the pyramids on the Giza Plateau.

The drilled holes are at predictable locations based on our theory. The solution might be as simple as going to the exact locations we expect evidence to be and digging to find the old marks.

Even before we visit the Giza plateau, aerial pictures provide us with valuable clues. A close examination of the pyramids shown above indicates that on each of the four sides of a pyramid, there is a line that runs from the cap at the top, straight down the middle of the side to the bottom. With proper sun lighting, it looks quite obvious. Some people have viewed this configuration and concluded that the pyramid has 8 sides.

In reality, this seems to prove that the construction was executed by independent crews, and where the work of two crews met, there was a seam that forms the line we observe on the pyramids. There is no overlapping of stone blocks from one crew side to the other. This is a very strong clue, which we hope to confirm during our documentation.

The other significant clues should be located at the corners of the super square. These can be located by going to the exact center of each pyramid side and establishing a perpendicular line to the side of the pyramid. If we measure a distance exactly equal to the length of the pyramid side, we will find the footprint of a post that was placed at each of the four corners of the bigger square.

The second source of evidence should be located along each side of the outer square. That should be making one-third of the square side. These markers were important for defining the size of the inner square.

The third source of evidence for the super square method is the footprints of the foundations of the four towers that were used to lift rock blocks up to different levels as the pyramid progressed upward.

The fourth source of evidence will be found along the sides of the pyramid. There should be regularly spaced drilled holes; these holes were the extensive foundation of the scaffolding. Scaffolding was required so the workers could climb to different levels of the finished part of the pyramid.

The scaffolding was connected to the towers for added support and stability and was anchored onto the surface of the pyramid. To provide an idea of the immensity of the work, the towers had to be slightly taller than the pyramid height. For the Khufu pyramid, that means 482 feet. The width of the pyramid was 722 feet at the bottom on each side, and all four sides were covered by scaffolding.

The construction site itself must have looked spectacular. For each corner workstation, there was a cut rock marshaling area and a staging area for preparing the rock blocks to be lifted. Once the workday started, the crews worked steadily lifting the scheduled number of blocks per hour or per day without interruption.

A close examination of the designated work station areas will yield clues as to the effects of the heavy traffic. For instance, there might be signs of rock ware in places where bedrock was exposed to wear and tear. Calculations will be executed to determine how many rocks were moved per crew per hour or per day in order to meet the completion schedule.

An examination of the size of the tower footprints will show the sizes of what the load was planned for. It is likely that the lower ends of the towers were built with larger timber pieces. It will also be possible to tell if there were secondary structures that were built to buttress the towers and make them stronger.

The vision of this book is to confirm that indeed the super square method was used to build the pyramids. When it is completed, the vision is to undertake a reality show, where we will capture the truth in real-time at the Giza plateau. We will use the Khufu pyramid for verifying the method, but we may also use either the Menhous or the Khufu pyramids to cross-check the consistency of the proofs.

Hopefully, the proceeds from the book will be able to monetize the documentary production. If not, alternative means will be sought. The excitement in this is that it has taken so long to produce a provable and reproducible pyramid construction methodology.

The limitations for providing a method are many. But the most significant one is time passage. After several millennia, evidence is lost. Subsequent human activity around the site also serves to obliterate the evidence. Then there is the issue of scope and size of the subject matter. It is difficult to choose where to start.

The builders too had a mission to befuddle their audience. They must now be chucking in their graves wondering why we cannot figure out what to them may have seemed like an easy puzzle. But their strategy was right. They had to make us think a little. No one would have been keen to investigate not just the constructed work but to critically discover what types of people they were. It would be interesting to interrogate them and discover all their secrets, but alas that is not possible. We have to settle for interpreting the few clues we can get our hands on.

CHAPTER 15

Plan for a Documentary

During the verification of the super square methodology, a documentary will be produced. It is appropriate to envisage what challenges might be involved in that endeavor. It is best to be prepared early rather than scramble later to organize resources.

The one outstanding thing is that the author recognizes pyramids as Holy places. He will not enter one without the authorization of spiritual leaders. These are different from the ministry of antiquities and other government authorities. In this era, it is difficult to find authorities that might be authentically aligned with the ancient beliefs, but there might be local guides who might be of assistance.

In this chapter, we will discuss a plan for initiating the ground verification of the super square method. Any work to verify hypotheses regarding Egyptian antiquities required express permission from the Egyptian authorities.

In this section, we will hypothetically explore the steps we will take to establish the truth on the ground. The stages should go as follows:

1. Submit the book for review by the Egyptian authorities (i.e., Ministry of Antiquities).
2. Request formal permission to conduct investigations.
3. Find sponsors.
4. Find partners.
5. Plan the physical visit to Egypt.
6. Arrange for video recording of the work.
7. Secure permission to be at the site for two months.
8. Conduct ground radar penetrating surveys of locations where we expect to see the evidence.
9. Obtain aerial photos of the target pyramid and secure photo images, including radar images.
10. Examine the bedrock at suspect locations for wear and tear.
11. Seek permission if necessary to dig for evidence at a few samples' locations.
12. Map findings providing precise measurements
13. Document all data collection precisely.
14. Write a report.
15. Submit a copy of the report to Egyptian authorities.
16. Publish report.
17. Release documentary.

The Egyptian authorities will have full control of our activities and will be fully informed of all progress. It is important that their culture and customs be respected and considered in whatever we do.

To be successful, the verification expedition needs to be organized starting one year prior to execution. Contacts with Egyptian authorities are the first priority. Permits or letters of authorization must be secured. Bookings of visits to the target area should be secured, preferably at low traffic times. Expertise must be sought and if necessary enrolled.

Six months prior to consultation with Egyptian Authorities, we will search for sponsors. The target agencies will be people with an interest in African history, public media, and historians. A list of potential contacts will be compiled.

Some agencies with expertise or useful resources may wish to join us as partners. These agencies would share both costs and proceeds if we obtain revenue from the endeavor. Once again, we will be looking for like-minded groups and potentially sympathetic groups to join us.

The physical visit to the target site will mark the beginning of the field project. For this to be successful we will need to compile a list of what equipment we will need to extract study data. We may need consultants to scope this out. People with prior experience with this type of venture will be consulted.

Generally, we have ideas of where to go to find evidence. What we need the most is the right equipment to extract the right data. Radar signal producers and sensors will be very important for this work. All efforts will be made to reduce physical disturbance of the site as much as possible. Remote imagers will be the preferred tools.

The data collection will mostly be by electronic means. By this, we mean that the data will be entered into computers based on predetermined forms. A database will be designed to collect specific attributes and to be saved in a specific format.

Maps of locations where data will be collected will be produced and information collected will be linked to a data location. Impressions of either unseen markings or hole drilling will be produced. Those should show a true picture of what lies under the surface outside the pyramid. We believe it will be an interesting untold story.

If our assumptions are true, the story of the history of pyramid construction will be quite compelling. For that reason, we expect the documentary to become quite popular.

We will need to undertake promotion of the new story. Once again we will need partners and promoters to assist us to get the story around. History is not a particularly popular subject for most people, however, the fact that pyramids are a popular destination for many tourists, we might find demand for our story.

There is no doubt that a better story of how Egyptians built pyramids needs to be told. Their contribution to our civilization is too important to be ignored. Not only that, the information modern civilization borrowed from those ancient times, must be acknowledged. Some valuable artifacts extracted from Egypt are circulating around Europe. Yet many Europeans have no respect for the blood and sweat that went into making them.

For example, there are huge support pillars taken from Egypt that stand in the Vatican. What credit is given to the Egyptians for producing them? How old are those pillars anyway? What value are they contributing to the Vatican? Similarly, many obelisks from Egypt are scattered in Europe. What value are they really to the Europeans?

Furthermore, should the Egyptians be seeking reparations for damages made to statues such as the Sphinx? Who is accountable for such damages? Should Egyptians be seeking to have such statues fixed? These are valuable historic relics that can bring pride to the African people. No one should take that history lightly, particularly considering that most developed countries continue to view Africans in a slavery context only. We need to hang onto any fragments of history that bring pride to the African people. Going back to the pyramids is like starting a new renascence of African history. The black people are tired of being downtrodden. In a sense, we need to take matters into our hands and tell our own stories.

CHAPTER 16

Restoring the Khufu Pyramid

This is a dream. I see myself walking the 21 kilometers barefoot from Cairo to the Khufu pyramid. And when I arrive there lo and behold the shining white marble wonder standing there bigger than life. It is a pilgrimage to the ingenuity of my ancestors. I cannot enter the pyramid because it is a holy temple. It was built to honor a god who up to now may be looking down on me and shedding tears for what we have done to his monument.

I reach near the front entrance of the pyramid and I pitch a tent there to observe his hand-work for a day. I spend a silent time there seeking his audience searching for communication from him to give me insight as to what the future holds. I don't come with demanding questions. I am there just to get a flow of information from him. How can we correct our path to the future? Our present is so bleak. The people are poor and getting poorer. The resources are dwindling. The climate is getting hotter. The rains are rare and far between.

Where should we seek solace? I ponder … has God abandoned us? Where did we go wrong? Are we beyond redemption? Will we achieve harmony, ever?

The restoration of at least one of the big Giza plateau pyramids is another potential reality show project. If we confirm the super square method as being what the Ancient Egyptians used to build pyramids, everything is possible. Restoration work is not as complicated as building the entire pyramid from the ground up. It would involve just the replacement of missing interior stone blocks and all the covers stones.

If one pyramid can be restored, then the restored structure can be on earth for another 20 thousand years. The unrestored pyramids will eventually turn to dust. The weathering process is eating away the internal rocks at present. With global warming and pollution, the existing structures may disappear in another two to three millennia. It would be dreadful if no effort is made to save at least one of them.

This is a construction project, not a documentation exercise. As such, much more planning work is required. It is also not a reproduction project. We do not have to do exactly what the Egyptians did five to ten millennia ago. The method verification work will allow us to prove that fairly advanced methods and tools were used to do the work. Reproducing what they did many millennia ago would be prohibitively expensive. This would particularly be the case if one insisted on using timber scaffolding. Timber supply is very limited in the Middle East.

What is required is a blend of modern and old techniques. Details cannot be provided here, but considerable consultation with engineers and experts will be undertaken.

The first step in restoring the pyramid would be to do a cost-benefit analysis of different options. The most efficient and cost-effective option would be selected.

Secondly, we would need to scope out how many inside cut stones are required. Surprisingly, not too many inside stones might need replacement. Probably less than 10,000 for the Khufu pyramid. The outer cover stones are the big problem. For starters, they have to be limestone or white quartz. For that reason, they might be difficult to find. The total number should be less than 80,000.

The other challenge is to find workers that can cut that many stones in a short time. You don't want to take years cutting stones. I would estimate a maximum of 4 months to cut stones. I assume modern equipment can be used to achieve that.

In this attempt at restoration, it might be wise to use a strong adhesive to get the cover stones to stick to the inner stones. Without the adhesive, water sips into the joint areas weakening the arrangement. That would be prevented if possible.

The system for placing the rock need not be described in detail at this point. But it is clear, that the weights to be lifted are still huge. Still, the work is much simpler than placing duct or passageway linings, chamber ceilings, and other specialty rocks.

After determining how many new rock blocks will be required, assessing what equipment will be required and what kinds of special skills will be required, then the true cost of the project can be estimated. It is apparent, that this could be a costly venture. But then, what cost can we attach to the sentimental value of the pyramids?

There may be many sympathizers who might be willing to commit to this achievement. It is a noble cause. And those that commit to it will get considerable satisfaction seeing the completed project. It will be a unique achievement never tried before on ancient pyramids.

Once again considerable effort will be required to enroll the Egyptian government on the benefits of a restoration. If the project is approved, permits and consent permissions will be required to close the subject pyramid for four to six months as the work is carried out.

To reduce the loss of revenue to the government, construction work should occur at a time when tourist traffic is lowest.

Of the three initiatives contemplated in this book, restoring the Khufu pyramid is the most difficult. It requires more resources, more rigorous planning, and the enrolment of a wider range of patrons, friends, and sympathizers to execute.

It is obviously doable, but our biggest human failing is a tendency to be skeptical at the slightest hint of technical difficulty. If it was done 5 millennia ago, it can be restored today. The amount of work required is less than 1% of what was done five millennia ago. And the generation of restorers claims to be much more advanced. So, there should be no skepticism.

Of course, there are some challenges that must be overcome, and they include; choosing a method for lifting the rock blocks. We have to use contemporary methods. Here the methods may include conveyor belt systems, cranes, pulleys, and others that may cop up. Hopefully, we can choose the most efficient methodology.

www.ingramcontent.com/pod-product-compliance
Lightning Source LLC
Chambersburg PA
CBHW050009040726
47599CB00014B/1285